ISDN
Networking Essentials

ISDN
Networking Essentials

Ed Tittel
Steve James

AP Professional

Boston San Diego New York
London Sydney Tokyo Toronto

This book is printed on acid-free paper. ∞

Copyright © 1996 by Academic Press, Inc.

All rights reserved.
No part of this publication may be reproduced or
transmitted in any form or by any means, electronic
or mechanical, including photocopy, recording, or
any information storage and retrieval system, without
permission in writing from the publisher.

All brand names and product names mentioned in this book
are trademarks or registered trademarks of their respective companies.

AP PROFESSIONAL
1300 Boylston St., Chestnut Hill, MA 02167

An Imprint of ACADEMIC PRESS, INC.
A Division of HARCOURT BRACE & COMPANY

United Kingdom Edition published by
ACADEMIC PRESS LIMITED
24–28 Oval Road, London NW1 7DX

Tittel, Ed. and James, Steve
   ISDN Networking Essentials / Ed Tittel
      p. cm.
   Includes bibliographical references and index.
   ISBN 0-12-691392-7
   1. Integrated services digital networks. I. Title.
   TK5103.75T58 1995
     004.6'6--dc20   95-20919
                                                                      CIP

Printed in the United States of America
  95   96   97   98   99   IP   9 8 7 6 5 4 3 2 1

Contents

About the Authors   ix

Acknowledgments   xi

Preface   xv
>   Introducing the Essentials Series   xv • ISDN Access Essentials   xvi • The Other Members of the Essentials Cast   xvi • Tell Us What You Think   xvii

Introduction   1
>   What Is ISDN?   1 • What Can You Do with ISDN?   2 • What Do You Need to Use ISDN?   3 • What's ISDN Going to Cost?   5 • Can You Do It Yourself?   6 • About This Book   7

1   ISDN Technical Overview   9

1 Digital Communications and ISDN 11

Communications Basics 12 • Digital Communications 14
Evolution of the Integrated Digital Network System 15 • ISDN Standards 19 • A Visualization of ISDN 21 • Summary 21

2 ISDN Structures and Functions 23

Transmission Structures 23 • User–Network Interfaces 27 • ISDN Protocols 29 • ISDN Connections 31 • Addressing 32 • Interworking 33 • Summary 34

3 ISDN Communications Service Links 37

T1 Networks 38 • Private Branch Exchanges (PBXs) 40 • Local Area Networks (LANs) 43 • TCP/IP (Internet) 45 • Summary 46

4 ISDN Technical Specifications 49

ISDN Protocol Architecture 49 • Physical Layer Protocols 51 • The D-Channel Data Link Protocol 59 • The D-Channel Layer 3 Protocol 66 • Packet- and Frame-Mode Bearer Services 71 • Asynchronous Transfer Mode (ATM) 74 • Signaling System No. 7 78 • Broadband B-ISDN 86 • Summary 90

II Current PC ISDN Hardware and Software 93

5 Determining Your ISDN Needs 95

So You're Serious About ISDN? 95 • How Do You Decide? 96 • What Do You Need? 104 • What Are Your Choices? 105 • Summary 106

6 Internal ISDN PC Adapter Cards 107

ISDN Internal Adapter Cards 107 • Summary 127

Contents

7   ISDN Modems and NT1s   129
    ISDN "Modems"   129 • Network Termination 1000 Devices (NT1)   131 • Meet the ISDN Modems   131 • Summary   152

8   Ethernet ISDN Bridges and Routers   153
    What Are Ethernet ISDN Bridges and Routers?   153 • ISDN Bridges and Routers with Ethernet Interfaces   154 • Summary   164

9   ISDN Software   167
    An Overview of ISDN Software Requirements   167 • ISDN Shareware/Free Software   170 • Commercial ISDN Software   174 • Summary   185

III   Getting Started with ISDN   187

10 ISDN Telephones and Business Equipment   189
    ISDN Telephones   190 • PRI   192 • Video Conferencing   197 • ISDN via Wireless and Satellite Equipment   199 • ISDN Test Equipment   201 • Summary   202

11 Making Good ISDN Choices   203
    The "Short List" of ISDN Acronyms   204 • Do It Yourself or Hire a Consultant?   204 • Hiring an ISDN Consultant   205 • Checklists of Uses, Hardware, Current Phone System, etc.   206 • Doing ISDN Yourself   208 • Summary   214

12 Getting Your Own Home ISDN System   215
    Ordering ISDN Service from Your Local ISDN Dialtone Provider   216 • Purchasing Your ISDN TA/NT1 Interface Card   217 • Ordering Your ISDN Internet Service   217 • ISDN Dialtone Service—Wiring, Installation, Configuration, and Testing   218 • ISDN TA/NT1 Interface Card Installation and Testing   221 • ISDN and Your ISP   228 • Summary   230

13 Individual Office ISDN Systems   231
Ordering the ISDN Service From Your Local ISDN Dialtone Provider   232 • Purchasing Your ISDN TA/NT1 Interface Card   233 • Ordering Your ISDN Internet Service   234 • ISDN Dialtone Service: Wiring, Installation, Configuration, and Testing   235 • ISDN TA and NT1 Installation and Testing   236 • Installing TCP/IP   239 • ISDN and Your ISP   239 • Summary   240

14 Troubleshooting, Tips, and Testing   243
Troubleshooting Your ISDN System   243 • Tips for Better ISDN Use   247 • ISDN Line Testing Equipment   249 • Summary   250

15 Frequently Asked Questions and Answers   251

A ISDN Dial-Tone Service Providers   263

B Vendor Information   267

C ISDN Bibliography and On-line Resources   275
What Does "On-line" Really Mean?   276 • CIS: The CompuServe Information Service   278 • The Internet   281 • Other Online Resources   286 • Summary   287 • Bibliographies and Resources   288

D Glossary   291

Index   299

About the Authors

Ed Tittel

Mr. Tittel is the author of numerous books about computing and a columnist for Maximize! magazine. He's the co-author (with Bob LeVitus) of three best-selling books: *Stupid DOS Tricks*, *Stupid Windows Tricks*, and *Stupid Beyond Belief DOS Tricks*. He's also a co-author (with Deni Connor and Earl Follis) of the best-selling *NetWare for Dummies*, now in its second edition. These days, he's turning his focus to Internet-related topics and activities, both as a writer and as a member of the NetWorld + Interop program committee. In that vein, he co-authored *HTML for Dummies* with the co-author of this book, Steve James, and *The Foundations of WWW Programming*, with HTML and CGI with Mark Gaither, Sebastian Hassinger, and Mike Erwin, both for IDG Books.

Ed's last "real job" was as the director of technical marketing for Novell, Inc. In this position, he tried his best to control technical content for Novell's corporate trade shows, marketing communications, and presentations. He has been a frequent speaker on LAN-related topics at industry events, and was even a course developer for Novell in San Jose, where he designed and maintained several introductory LAN training classes.

Ed has been a regular contributor to the computer trade press since 1987, and he has written more than 150 articles for a variety of publications, with a decided emphasis on networking technology. These publications include *Computerworld*, *InfoWorld*, *LAN Times*, *LAN Magazine*, *BYTE*, *Macworld*,

MacUser, NetGuide and *IWAY*. He is also a regular columnist and contributing editor for *MAXIMIZE!* magazine.

You can contact him at
512-452-5670
512-452-8018 (FAX)
CompuServe ID: 76376,606/Internet e-mail: etittel@zilker.net
<URL: http://www.io.com/~mcintyre/lanwrghts/lanwrghts.html>

STEVE JAMES

Mr. James is a long time computer-industry writer who's covered the documentation needs of organizations as diverse as the U.S. Army Corps of Engineers and The Psychological Corporation. A former biological researcher, Steve has concentrated his efforts in some computer-related operation or another for the past 15 years. Along the way, he's fathered more than 50-odd manuals and other lengthy works of technical prose, and made some excellent friends along the way. He's also the co-author, with Ed Tittel, of *HTML for Dummies*.

Currently, Steve divides his time among the keyboard, his family, and the great outdoors, where the thrill of competitive bicycling continues to lure him, despite his accelerating decrepitude.

You can reach Steve on the Internet at snjames@wetlands.com.

ACKNOWLEDGMENTS

ED TITTEL

I have an awful lot of people to thank for making this book possible, and for contributing so substantially to its contents and direction. First and foremost, I must thank the people who helped me research, write, and assemble this book. These individuals include:

- **Dawn Rader**. Dawn is the managing editor at *NetWare Solutions* Magazine, where she helps keep a close watch on the chaotic and intense activities involved in getting a magazine out the door every month. Somehow, she found the time to do that, get married, handle the copy edits for this manuscript, and more!
- **Michael Stewart**. Michael's been my chief researcher, right-hand man, WebMaster, and equipment guru for quite a while now. All I can say is that I've come to depend on him for the many services, large and small, that he provides so cheerfully and reliably. Thanks for being there when I needed you, Michael!
- **Steve James**. Steve is both co-author and colleague, who struggled mightily to install and debug numerous ISDN widgets, devices, NICs, and who took quite a few knocks along the way! Sorry for all the bike rides you missed, but it was fun learning and working together.

- **Dave Smith**. Dave is a man of many more talents than I've been able to figure out how to use. He backed me up as an editor and writer on this book, and assisted with the overview chapters in Part I. I can't say enough good things about his willingness to learn and his writing ability, so I'll stop with a giant: "Thanks again! Dave!"

With this, the fifth book that we've done together for AP PROFESSIONAL, I'd also like to continue my ongoing thanks to Susan Price (and to her faithful associate Michelle Ryan) for doing such a good job with our sometimes scattered materials. Finally, I'd like to thank the technical editor Lenny Tropiano, for picking as many nits and issues as he could find, and generally improving the value and accuracy of the book.

There are also lots of folks at AP PROFESSIONAL who deserve thanks and mention. First and foremost, I'd like to thank my project editor, Jenifer Niles, for sticking with me during a period when one delay seemed to lead inevitably to another. Once we got the "final, final, final" deadline straightened out, I'm pleased to say that everything else went without a hitch!. Thanks also to Jacqui for managing the many logistical details and minutiae of pulling a big book together.

Also, I'd like to thank the many vendors and networking experts whose hard work and good advice made this book possible. If it weren't for the work all these people had done before Steve, Dave and I got started, we couldn't have marshalled the fact and figures you'll find in here. The many individuals and companies who provided us information, evaluations, referrences, and resources are just too numerous to mention by name, but I'd like to wish them all a fervent and well-meant "blanket thank-you" all the same. However, we reserve the right to thank Dan Kegel in person for putting together the best online resource on ISDN that there ever was (or will be)!

Finally, I'd like to thank my family for putting up with me through the course of a time-consuming and all-absorbing project. I've learned that being a "virtual husband" and a "virtual stepfather" is nowhere near as much fun as the real thing. So, Suzy, Austin, Chelsea, and Dusty, please accept my thanks for sticking with me, and my apologies for the many days and nights I was unavailable because of this book!

Ed Tittel, Austin, TX, September 14, 1995.

Acknowledgments

STEVE JAMES

First and foremost, I want to thank Ed Tittel for asking me to co-author this book with him. He has been my inspiration and mentor in this exciting adventure into ISDN. My eternal gratitude goes to my understanding family, Trisha, Kelly, and Chris, for putting up with my writer's quirks and schedule, especially during late-night work sessions.

I greatly appreciate the kindness of George Wenzel, of RealTime Communications, who took time out of his busy schedule to enlighten me about the inner workings of ISDN from the Internet service provider's point of view. Finally, I'd like to thank the folks at Southwestern Bell Telephone for keeping my ISDN line up throughout the researching and writing, and all of the people who posted their wonderful replies to the myriad of questions on the comp.dcom.isdn newsgroup.

Steve James, Austin, TX, September 14, 1995

PREFACE

INTRODUCING THE ESSENTIALS SERIES

Welcome to networking! You may be new to the topic, or have some experience with computer networks; either way, we want to help you learn more. In this volume of the series, we tackle the sometimes unavailable, and always mysterious wide-area networking technology known as the Integrated Services Digital Network, or ISDN.

Maybe you've already looked for a reference on ISDN and have been overwhelmed by engineering and technical content, but underwhelmed by the lack of hands-on discussion or practical applications. According to our own bookstore survey, there are about 10 titles on ISDN in print already. So why do we, your humble authors, need to write another one? The answer's in the concept behind the series, and in how it informs this book.

The *Essentials* series aims to fill the gap between theory and practice for networking. It offers solid, practical, networking information, without jargon or unnecessary detail. It explains all its topics in clear, concise language, and plainly defines terms that may be unfamiliar.

The series is divided into the topics that are most likely to be of use to networking professionals or serious users, from setting up a network, to finding problems with it, to more specialized topics such as mobile computing, ISDN, and electronic messaging. Each book takes a gentle, light-hearted approach to the material that focuses on the basic principles you absolutely must know to master a topic or technology, with plenty of tips and proven techniques gleaned from the authors' wealth of practical experience in putting theory into practice.

ISDN Access Essentials

Where ISDN is concerned, what makes our book different from some of the others is that we assume little or no foreknowledge of computer telephony, digital signaling, or B-ISDN, and lead you into each topic as best we can. In fact, we've tried to lay out the materials so there's something for everyone, whether beginner, novice, or seasoned cybernaut.

What makes our book different from some of the other ISDN tomes is that we don't attempt to provide an encyclopedic reference to all conceivable ISDN protocols, services, and abilities: Our book is designed as much to provide a set of techniques for you to do your own investigation and installation, as it is to provide a roadmap to what's involved in using and working with ISDN. Finally, our book is designed to provide useful tools and techniques to help you learn for yourself, and to supply you with the essential information you'll need to be able to navigate and cope with the many viewpoints, implementations, and discussions you're likely to find surrounding ISDN.

The Other Members of the Essentials Cast

The first two books in this series will be of interest to anyone operating a local area network. The first, *Network Design Essentials*, covers everything the reader needs to know to put a smoothly running network in place, including the maintenance tasks that will keep it running. The second book, *Electronic Mail Essentials*, is intended to be useful to anyone who uses message-based applications over a network, or who needs to understand how to design, install, or manage an electronic mail system.

The remaining books in the series are intended to expand on networking topics that should apply to most, but not all, network administrators. This book is aimed at helping PC users who wish to use ISDN on standalone machines, or on small networks, get the equipment, software, and information they need to establish a working connection in short order.

The third book in the *Essentials* series responds to the recent explosive growth of that huge network of networks, the Internet, and is designed to help you to tap into the enormous and sometimes overwhelming treasure trove of information that the Internet represents. And if your computing needs extend to large, far-flung networks, you'll definitely want to investigate both *Remote Access Essentials* and *Computer Telephony Essentials*, which should appear in bookstores in late 1995 or early 1996.

Tell Us What You Think

The authors want to hear from you about these books. Please write to us, care of the publisher, or contact us through the electronic mail addresses we provided in the About the Authors section. We'd like to hear more about what you liked and didn't like about these books, and on your ideas for other topics or more details that you need for your networking life.

In closing, we'd like to say: "Thanks for buying this book." We hope you find it worth the money you've paid for it, and the time it takes to read!

Introduction

Welcome to *ISDN Access Essentials*! This book is aimed squarely at the following audiences:

- People who are curious about how ISDN works and what it can do for them at a personal or small-business level
- People in small or home offices with PCs who want to establish a working ISDN connection, probably to improve the performance of their Internet link
- Network administrators or power users on small networks (15 nodes or less) that want to link their users to the Internet or another information provider on a network basis

What Is ISDN?

ISDN stands for Integrated Services Digital Network. While most of this book is devoted to answering the question posed in the heading for this section, the quick answer (to help you get started) includes the following elements:

- A high-speed way of connecting computers or networks together
- A purely digital means of communication that uses the existing telephone network
- A way of combining voice, data, video, and more over the same communications line
- A way for the telephone companies to leverage their existing investments in digital technology, and make more money off their telephone lines
- A faster replacement for data communications commonly served by normal telephone lines and modems today

Actually, ISDN is all of these things, and more, as we'll try to inform you throughout this book. But its best advantages, by far, come from the fact that ISDN is a purely digital form of communication, all the way from sender to receiver. This lets ISDN handle more data more cleanly and effectively than existing analog telephone technology, while also opening the door for more advanced digital forms of communication (such as full-motion or free-frame video, multimedia, interactive whiteboards, and other "far-out" applications).

What Can You Do with ISDN?

Despite the great potential for advanced video and multimedia applications that ISDN may support some day, ISDN today represents a cleaner, faster, and more feature-rich replacement for current telephone and telecommunications technology. Because the new wave of applications isn't far enough along to make video teleconferencing or picture-phones consumer products, these capabilities are still more technological curiosities than main lines of business.

Although the range of services and capabilities that ISDN delivers will change dramatically in the next decade, ISDN delivers the following services and primary characteristics today:

- Advanced voice call features, including call-waiting up to 15 levels deep, automatic call-forwarding and call-following services, multi-party conferencing, etc.
- From 64 Kbps to 128 Kbps data communications capabilities, at costs significantly lower than existing fractional T-1 or "switched 56" lines
- The ability to integrate voice and data communications for one or two data lines and one or two voice lines (usually one or two data and a single voice line) for the price of a single connection
- Significant performance improvements over conventional modem communications, even considering the theoretical 115 Kbps throughput sometimes available from V.34 modems using V.42bis compression

We think these things make ISDN worth considering immediately, even discounting the fabulous future that ISDN appears poised to deliver some time in the next decade. For us, in fact, the ability to speed our Internet access by an order of magnitude was the primary attraction that led us down the path to a successful ISDN installation in the first place.

What Do You Need to Use ISDN?

Today, ISDN is available to about 35% of the computer-owning households and businesses in the United States, but by 1997 that figure should climb to over 50%. To participate in the potential bounties of ISDN services and capabilities, you'll need to make sure that service is available in your area, and then you'll have to obtain the necessary ingredients to establish a working ISDN connection.

Here's the short list of what's necessary to bring ISDN into your home or office:

- You'll have to determine that ISDN service can be delivered to your home or office. The best way to find out if it's available in your area is to call your local phone company. Ask to speak to their business service department (even if you're thinking about a home installation), and then inquire about the availability of ISDN service. If the response isn't satisfactory (i.e. nobody knows if it's available or not), call the phone companies headquarters location and ask them instead. ISDN is still new enough that only a select cadre within the phone company can be considered "ISDN-friendly" if not "ISDN-knowledgeable."

- Assuming that ISDN service is available (right now, the odds are 1 in 3; actually, they're greater if you're in a metropolitan area, and much less if you're ensconced in a rural setting), you'll need to schedule an ISDN installation. Before you do that, be sure to ask how much this will cost you, and if any special promotional deals or considerations are available. If you make the commitment, ask for a date for installation (in many areas a four to six week wait is not at all unusual).

- Once you know the date for your installation, you can schedule the remaining activities around it. You'll want to arrange to have cables installed from your phone company's demarcation point ("demarc") on your premises to wherever you want to attach your ISDN equipment. You can install this wiring yourself, but we recommend hiring a qualified cable installer to do the job. Look in the Yellow Pages under Telephone Installers or Computer Wiring and Installation to find prospective candidates for the job. Ask for two or three bids, and take the one that combines the best price with the best reputation and service.

- You'll also need to decide what kind of ISDN hardware to buy: normally, this includes either an internal or external ISDN adapter for your PC, plus a network termination device. We devote three chapters in this book to aid you in this selection process. Once you've made your decision, you'll need to order the necessary components and obtain the requisite software to make things work.

- Once ISDN service is installed, make sure you get the right information about your connection from the phone company. In addition to various line identifiers (SPIDs), you'll also need to find out what kind of telephone switch your ISDN connection is routed through, and how it has been set up for your use. It will also be helpful to tell the phone company what kind of hardware you plan to use, so that you can both be reasonably sure that things will work together properly.

- Finally, you can install your PC hardware and software, and go through the process of getting things working. If you're lucky, this

Introduction

will be a matter of a day or two; if you're not, it could take two weeks or longer to work out all the kinks necessary to get things up and running. Be prepared, be calm, and be patient—this can be a real roller-coaster ride!

Once you've survived the rigors of the installation process, you should be ready for just about anything! That's when the real fun begins, anyway.

What's ISDN Going to Cost?

As more and more participants join the fray, prices for hardware and connections should drop somewhat, but today ISDN is not exactly cheap. Here's what a typical single-user ISDN installation can cost in the United States:

Item	Charge	Explanation
Upfront charges		
Installation fee	$250–500	Waived in some areas, special promotions sometimes available.
Wiring charges	$50–200	Additional on-premises wiring required to bring ISDN to your computer
Hardware expense	$450–1000	You'll need ISDN hardware for your PC, and a terminal adapter to connect to ISDN.
Monthly fees		
ISDN service	$40–80	Line charges for your ISDN line.
Internet access	$80–150	ISDN access charges from your Internet Service Provider.

Costs will vary from area to area, because different ISDN service providers (usually your local telephone company) have different installation charges, and monthly fees can vary from a low of $40 (including taxes and fees) to a high of $80 per month.

Basically, you're looking at spending between $750 and $1,700 to establish your ISDN connection, and between $120 and $230 a month (or more) for recurring charges. Equipment costs aside, this compares favorably with the cost of a business line in most parts of the U.S., but is considerably more expensive than residential phone service.

CAN YOU DO IT YOURSELF?

By this point, you've probably gotten the idea that (a) ISDN is kind of expensive and (b) that it's not the simplest connection technology in the world to install and get working. Both of these points are true, and should lead you to another question—namely, "Can I do this myself?" If you're smart, you might even have phrased the question as "Do I want to do this myself?"

Although your budget will have a strong bearing on the answer, we're inclined to suggest that you should consider hiring a consultant to help with the installation, no matter what your financial circumstances might be. Of course, you can still decide to go it alone, but it's worth considering that you may have to devote significant amounts of time and effort to get ISDN up and running.

We need to warn you plainly that doing it yourself might mean spending a "mere" 20–30 hours on the task, including researching the hardware selection, arranging for installations, and going through the installation on your PC. But it could also mean spending 40, 60, 80 hours or more on the phone with your ISDN provider, your Internet service provider, the NT-1 vendor, the ISDN interface vendor, and your communications software vendor as you try to puzzle out why your chosen combination of components isn't working.

Therefore, if your time has value, you also need to consider the potential "opportunity costs" of being an ISDN do-it-yourselfer. Our advice is that if 30 hours of your time is worth more to you than the costs of hiring a consultant for your installation, that you locate and hire a consultant without giving the matter much further thought. If you can't afford the costs, or don't want to spend the money, be prepared to spend some significant time and effort in getting your ISDN installation up and running!

About This Book

ISDN Access Essentials has been formulated with four primary goals in mind:

1. To help you understand enough of the workings, costs, and capabilities of ISDN to decide if you want to use it or not.
2. To provide you with sufficient information about available ISDN hardware and software offerings to equip you to make informed purchase decisions.
3. To supply you with sufficient background and terminology to be able to survive the rigors of the installation process.
4. To acquaint you with the best sources for up-to-the-minute ISDN information so that you can temper whites included in this book with the latest and greatest technical and product information.

To that end, this book has been structured into three parts, with numerous appendices, to provide the information you'll need to reach each of these four goals:

1. Part I of the book introduces the fundamentals of ISDN technology and terminology, and explains what it is and how it works.
2. Part II of the book introduces current ISDN hardware and software options, with rating and pricing information to help you select the items appropriate for your needs and your budget (Appendix A lists ISDN service providers to help you locate one in your area, while Appendix B includes comprehensive listing of ISDN vendors).
3. Part III of the book covers the installation and troubleshooting of a typical PC ISDN setup, to acquaint you with the process, and to equip you to deal with the kinds of problems or questions you might encounter along the way.
4. The remaining appendices provide bibliographic and on-line resources to help you stay current on ISDN technology and products, including pointers to the very best resources on ISDN, whether in print or in electronic form.

We suggest that you read this book from start to finish if you're relatively inexperienced with ISDN. If you're already somewhat knowledgeable, you can probably skip Part I, and delve into those areas of Parts II and III that interest you. We also suggest that all of our readers sample the various appendices, if only to determine what kind of information they contain, and to see if their contents might provide some useful pointers to vendors, consultants, or other sources of ISDN enlightenment.

Whatever your approach to this book, we hope you find it useful. Please feel free to correspond via e-mail with either of the authors. We welcome your feedback, whether critical or otherwise. And we hope that a better second edition can result from the input from our readers!

part I

ISDN Technical Overview

The Integrated Services Digital Network has been the topic of fear, hype, and speculation for years now. But it's only since early to mid-1994 that there's been any realistic hope of having that technology delivered to your home or office in a usable way.

For years, this led techno-humorists to decode the ISDN acronym as "I Still Don't kNow" or "It's Still a Dream, No?" But today, the only problem in obtaining ISDN is a lack of universal coverage. Virtually all of the former "Baby Bells"—the Regional Bell Operating Companies (RBOCs)—are offering ISDN to their customers, at least in pilot implementations, and other carriers, such as MCI and Sprint, are entering this business as well. Although universal ISDN coverage is as much as a decade away in the US, ISDN is no longer a pipe dream.

In this, the first Part of the book, we introduce the underlying terms and technologies that make up and explain ISDN. Starting in Chapter 1, we begin with a discussion of the basics of digital and analog communications, and explain what makes ISDN so interesting and special for computer-to-computer, and human-to-human communication. In Chapter 2, we lift ISDN's hood a bit to examine the transmission structures, interfaces, and protocols that make ISDN possible. Then, in Chapter 3, we discuss the various kind of private and public networks that can accommodate ISDN traffic, and what their relationships are. Finally, in Chapter 4, we take an in-depth look at the specifications for ISDN, including protocols, bearer services, signaling regimes, and more.

Our goal in Part I is to equip you to understand the terms and concepts behind ISDN so you can better understand the technology and its capabilities. Along the way, you'll find yourself learning more about the operation of the telephone system, both long- and short-haul, than you ever dreamed possible. We hope you enjoy the trip!

chapter 1

Digital Communications and ISDN

Nothing ever just happens. Circumstances, coincidences, and events converge; cultures combine, individuals act and react; or natural forces combine and recombine in patterns that have never existed before, or in differing patterns that have been replicated since the beginning of time. The natural world supports the occasional revolutionary change, but all change is essentially evolutionary when one takes the long view.

Since the origins of humanity there has been technology. The history of human civilization goes hand in hand with the history of technology. There is a logical and natural set of circumstances and perceived needs that has led us from the stone ax, the lever, and the wheel to electric guitars, in-the-door ice dispensers, and digital communications. What we call state-of-the-art technology has its roots back in the dream times of human history.

What do these ruminations have to do with the Integrated Systems Digital Network (ISDN)? Our point is that ISDN, while it appears to be a revolutionary way of handling communications, is the natural result of evolutionary processes that began back when digital communications meant holding up two fingers and pointing at a herd of mastodon.

Since the beginnings of human history, computing systems and communications systems have been evolving from the simple to the complex. While the importance of nonverbal digital communications in Stone Age hunting strategy should not be minimized, it is a subject somewhat outside the scope of this book. The intent of this chapter is to show how the Integrated Systems Digital Network is the result of and a synthesis of evolutionary processes in communications and information processing technology.

Communications Basics

ISDN really represents the inevitable convergence of two related technologies: communications and modern distributed computing. Each field grew up more or less independently of the other. As their development progressed, the benefits of their conjunction became more obvious, and more of a necessity.

Communications technology, most particularly telephone communications, has developed out of a perceived need to provide the benefits of face-to-face communications over long distances. The developers of the telephone set out to develop a technology, *telephony*, that would allow the sound of a human voice to be carried over long distances and recreated at a receiving station. In little more than 100 years, telephone technology has been refined to create the modern worldwide network that we take more or less for granted today.

Computer technology results from the perceived need to process and store volumes of information generated in normal human interactions. The need to solve problems, and to store and distribute information, has been common to humanity from the days of painting on cave walls up to present-day enterprise computing. The modern solution to this need has been the development and refinement of more flexible and more powerful computer technology. In a short 40 years, computer technology has evolved rapidly from the batch-processing, vacuum-tube mainframes chugging out jobs, to modern enterprise LANs and WANs, supporting state-of-the-art distributed systems.

Analog Communications

Telephone communication is based on human speech. The original goal of the telephone's developers was to provide a means to transport the human voice faithfully (or at least audibly) over long distances. In its beginning phases, digital telephony wasn't even a pipe dream. Early practitioners of the telephonic arts used the human voice, which they were determined to transmit and recreate, as the basis for development of their systems.

Human voices generate sounds that travel through the air in continuous waves. These waves are known as *analog* signals. Analog voice signals cause vibrations on a membrane in the human ear which converts the waves into recognizable sound. Not surprisingly, a conventional telephone receiver acts much like the human ear: Speaking into the phone transmits vibrations onto a membrane that generates electric impulses that are transmitted as a continuous wave over the phone lines. On the other end of a connection, telephone receivers interpret these electrical impulses and recreate corresponding sound waves as close to the original as the limits of the equipment will allow.

Since these electric pulses are analogous to the sound waves that generate them, telephony is said to be based on analog transmission, in the same way that human speech is based on analog transmission.

Computer Design

While the telephone network was reaching worldwide, the development of modern computing began. All communications inside computing machines, between computers and their peripheral devices, and between interconnected computers is handled digitally.

While the telephone system was originally based on analog transmission and evolved from that point, there has never been a precedent for analog transmission in a computing environment. Computers have always been based on digital information and instructions. Thus, communications inside the computer and among devices attached to computers has always been digital.

Construction of the Telephone Network

The original phone system was made up of a single transmitter connected to a single receiver over a single wire. Early commercial telephone implementations offered phone service only to sites directly connected by discrete wires. As the popularity of phone service grew exponentially, the basic components that now make up the international telephone network began to be developed and brought on line.

The first addition to the phone system was the addition of *central offices* (COs), which provided a common switching point for all the phone lines in a given geographical area. All telephone system end users were connected to the CO by their *subscriber loops*. A call would come through the central office over a subscriber loop and a connection to the receiving subscriber would be made. At first these connections were made manually, with operators physically connecting wires on patch panels to establish the connections between subscribers.

In the next phase of elaboration, COs were linked by trunk lines. Multiple COs could also be linked by *tandem offices*. A tandem office contains a *tandem switch* that can switch transmissions over trunk lines and handle the routing of calls between COs. Tandem switches lower the cost of phone transmission by providing links between COs that do not require a dedicated direct trunk line.

Since 1984 the United States has been divided into 161 *local access and transport areas* (LATAs), made up of local loops, COs, and tandem switches. Long-distance calls are handled by interLATA carriers. COs or tandem offices switch calls to the interLATA carriers, where calls are transmitted and switched between LATAs in the same way they are transmitted and switched between COs. This may sound complex (and in many ways it is), but it also means that almost any call in the U.S. can get between caller and callee by going through less than six involved parties: two local loops, two LATAs, and one or two interLATA carriers. Logically, this bespeaks a complex and pervasive communications grid.

Digital Communications

Computers have always used digital signaling methods to control local devices, and for communications over local buses. The development of the local area network (LAN) led to a slightly more complex form of digital communication—namely, communication between and among computers. LANs permitted powerful new applications for computer technology. In fact, many industry pundits argue that the development of LAN technology was an important first step toward a new widespread use of computers, the linking of LANs into wide-area networks (WANs).

Widespread adoption of WANs has led to large-scale use of the telephone network to complete the links between widely separated computers. Computer network nodes on a WAN may be connected not only by local cables, but also by telephone links.

Normally this requires that digital computer signals be converted to analog signals for transmission over the phone systems, and then reconverted on the receiving by modulating and demodulating equipment (modems). Even though this conversion causes considerable time delays and increases network overhead, the benefits of wide-area networking make it practical and necessary. Widespread use of telephone lines for data transfer created a new use for phone switching and transmission equipment and has been a key factor in the push for a fully digital telephone system.

EVOLUTION OF THE INTEGRATED DIGITAL NETWORK SYSTEM

The development of digital computer technology fomented a revolution in the way organizations created, stored, and transferred information. Many key innovations originally developed for computer technology were quickly adapted for use on the telephone network.

Analog to Digital (and back again)

Telephone networks depend on the coordination of many separate and distinct parts for their operation. Development of these parts has traditionally been concentrated in two major divisions: transmission and switching. Developments in computer technology have traditionally been adapted for use in the telephone network to lower cost and improve the quality and reliability of service.

The first major use of computer, or digital, technology in the phone network was in transmission. Although most local subscriber loops in the United States are still analog, the more sophisticated transmission functions of telephony have been converted to digital. The first digital lines used in the U.S. were *digital first carrier systems* (T1), made up of two wire pairs, one for transmission and one for reception. T1 lines were introduced into the phone network to provide a high-volume link between COs, and between COs and tandem offices. Analog signals are converted to a digital bitstream and transmitted over the T1 line, where they are reconverted to analog on the other end.

Analog Signal Transmission

Analog signals are continuous signals consisting of waves traveling in cycles. Human voice, video, and music are all examples of analog signals. The frequency of an analog signal is measured in hertz (Hz), or cycles per second. The *passband* of an analog channel is defined as the range of frequencies that can be carried simultaneously. The *bandwidth* is the width of the passband required for transmission. Different channels may use different passbands within the frequency range of transmission.

The passband of an analog telephone link is defined as approximately 300 to 3400 Hz. The human voice produces sounds between 50 and 1500 Hz. Obviously the passband of the physical phone link isn't adequate to carry the full range of human voice, but research has shown that the majority of human voice frequencies fall between 300 and 3400 Hz. Telephone links are optimized to carry these frequencies, which contain enough of the range of human sound to recreate clearly recognizable speech.

The bandwidth of voice communications is limited on a telephone network, so that multiple phone conversations can take place over the same physical channel, or link.

Digital Signal Transmission

For digital transmission over T1 lines, and to take advantage of digital switching, analog voice must be converted into a digital bitstream. The continuously varying values of an analog signal are sampled 8000 times per second and converted to digital values using a coding algorithm called µ-law encoding. A continuous set of eight bits carries this digital value. After the value is obtained it is mapped to one of 254 different numerical volume numbers (amplitude values).

Commonly an analog signal is generated, converted to a digital stream, transmitted, and then deconverted at the receiving end into analog signals. Ironically, the use of computer modems adds another set of conversions to the each end of the stream, where digital data is converted to analog for initial transmission over the local loop, and on reception from the local loop, converted back from analog into digital.

Multiplexing

Using a common link for many simultaneous connections is called *multiplexing*. Multiplexing in the telephone network allows for multiple conversations or connections occurring over the same physical connection.

Analog and digital signals are multiplexed in different ways. The analog telephone network uses *frequency division multiplexing* (FDM) to carry multiple conversations. FDM apportions the total available bandwidth into channels or bands that belong to an assigned user for the duration of the connection. For voice transmission, each conversation is assigned a different passband with a bandwidth of 3100 Hz.

Digital signals are multiplexed over a link using *time division multiplexing* (TDM). TDM gives each separate channel the entire frequency range for a very tiny increment of time, before switching to the next transmitter waiting in line. That's why TDM is sometimes called a "time-slicing" approach; nobody gets the whole bandwidth all of the time, but everybody gets a slice for their portion of each unit of time.

Switching

The first switching devices on the telephone network were electromechanical switches called *step-by-step* switches. As their name implies, step-by-step switches react to each digit dialed by the user and make physical connections in the switch to route the call. The next step in switch development was to provide *common control* of the switching function within the switch. Common control switching is based on a series of electromechanical relays, where instructions for handling the switching functions are statically defined by the internal wiring inside the switch.

Stored program switches were made possible by the development of the transistor. These switches allowed easier reprogramming of connections and call control within the phone network. The first stored program switches were installed in the early 1970s. Essentially, they were digital switches, so their inclusion into the telephone network, combined with the implementation of T1 lines, marks the beginning of digital telephony, and the beginning of the evolution of ISDN.

The inclusion of digital switches and high-volume digital transmission lines required a high volume of analog-to-digital and digital-to-analog conversions within the network. While these conversions were practical and necessary, considering the evolution of the phone network, they also added considerable overhead and cost to transmission of phone calls without adding any value to the connections. Engineers in the phone network immediately began to convert more of their transmissions and switching functions from analog to digital, in order to cope with this overhead and its associated costs by reducing the number of conversions required.

LANs and WANs

Meanwhile, the development and implementation of computer technology continued at an incredible pace. New ways of constructing and using computers were envisioned, developed, deployed, improved, and replaced with more efficient methods in a ceaseless quest for the best and fastest technologies.

One of the most significant changes in computing followed the move from batch-oriented mainframe processing to distributed computing. Following this chance in physical organization, innovations pushed computing toward decentralized processing and storage. Most of this push resulted from the deployment of local area networks composed of powerful desktop computers spread willy-nilly around many enterprises.

At the same, these newly deployed LANs were linked by telephone circuits to form wide-area networks (WANs) that vastly increased the desktop computer user's access to application processing, storage capacity, and sources of information.

The widespread implementation of LANs and WANs placed a new burden on telephone transmissions. Primarily designed and built to provide analog voice transmission, the telephone network was increasingly called on to provide the transmission of digital data.

As we mentioned earlier, digital computer signals are converted to analog signals (modulated) on the sending end of a transmission, and deconverted back to digital signal at the other end (demodulated). Modems (MOdulate/DEModulate) are designed to handle this task to enable computer communications over the telephone networks. But because the infrastructure of the telephone network was increasingly shifting from analog to digital, computers invariably had to go through an extra layer of modulation and demodulation, simply to use what was primarily a digital medium in the first place!

The Move to IDN

As more and more phone circuits are used for digital computer data transmission, new developments and evolution in digital hardware and software continue to bring down the transmission and switching costs and overhead necessary for the operation of the telephone network. From the beginning, it has been obvious for reasons of economy, expedience, and accuracy that conversion and deconversion of analog and digital signals over the telephone network should be eliminated.

As early as 1959, when the first experimental digital telephone technology was being tested and debugged, it was proposed that the global telephone network should move from its analog beginnings and be converted into an integrated digital network (IDN), where switching and transmission facilities in the network could be combined for efficiency.

Users and implementers of wide-area networks have always recognized the value and sought the implementation of end-to-end digital connections for transfer of computer information. Telephone network technicians sought out the cost savings, increased load capacity, and tighter control that can be provided by eliminating analog-to-digital conversions inside the system.

ISDN

The word "integrated," in IDN, implies the integration of switching and transmission. IDN has meaning in the context of the international telephone network and embodies the goal of the total digitalization of telephone technology.

The concept of ISDN is somewhat broader. In ISDN, "integrated" means that data transmission of many types—voice, video, sound, and data—may be seamlessly included in the digital switching and transmission capabilities of the IDN.

ISDN Standards

Standards are defined as "a prescribed set of rules, conditions, or requirements concerning definition of terms; classification of components; specification of materials, performance, or operation; delineation of procedures;

or measurement of quantity and quality in describing materials, product, systems, services, or practices."

Several standards organizations have involved themselves in defining the standards for the proposed ISDN. The preeminent body in this group is the International Telegraph and Telephone Consultative Committee (CCITT).

The CCITT definition of ISDN is as follows: "An ISDN is a network, in general evolving from a telephony IDN, that provides end-to-end digital connectivity to support a wide range of services, including voice and non-voice services, to which users have access by a limited set of standard multi-purpose user-network interfaces."

In 1984 the CCITT defined the following *principles* for ISDN:

1. ISDN should support a range of voice and non-voice applications. Service integration for an ISDN should take place using a limited set of connection types and user-network interface arrangements.
2. ISDNs support a variety of applications, including switched and non-switched connections. Switched connections should include both circuit-switched and packet-switched connections.
3. New services introduced into ISDN should be arranged to be compatible with 64 kbps switched digital connections.
4. ISDNs will contain intelligence for the purpose of providing service features, maintenance, and network management functions. This intelligence may not be sufficient for some new services and may have to be supplemented by either additional intelligence within the network, or possibly compatible intelligence in the user terminals.
5. A layered protocol structure should be used for the specification of the access to an ISDN. Access from a user to ISDN resources may vary depending upon the service required and upon the status of implementation of national ISDNs.
6. It is recognized that ISDNs may be implemented in a variety of configurations according to specific national situations.

Further, the CCITT recommends the following evolutionary pattern for the move to ISDN:

1. ISDNs will be based on concepts developed for telephone ISDNs, and may evolve by incorporating additional functions and network

features including those of any other dedicated networks such as circuit switching and packet switching for data so as to provide for existing and new services.
2. Arrangements must be developed for the interworking of services on ISDNs and services on other networks as transition from pre-ISDN to ISDN networks proceeds.
3. An evolving ISDN may also include at later stages switched connections at bit rates higher and lower than 64 kbps.

A Visualization of ISDN

The natural evolution of ISDN communication reached critical mass in the 1970s, when digital switches and transmission lines were introduced and started to become ubiquitous in the world telephone network. Evolution and rapid improvement in computer technology expanded the computer's role in communications while the IDN became a reality. Implementation of standards for the envisioned ISDN makes the union of the two technologies complete

Summary

The first worldwide network to establish interconnecting links was the telephone network. The telephone was developed to provide communications capabilities over long distances that were analogous to human speech.

As more and more users got into the phone network, various additions and improvements were made to allow for the increase in demand and keep the cost of services as low as possible. Original telephone operations were improved and linked up with the new technologies to create a network of telephone links that covered the entire world. Many components of the original analog system were replaced by digital equipment as it became available, making it possible for overhead and costs to be lowered.

Computers have always been based on communications using a stream of binary bits. In other words, computers have always used digital communications. At the end of the 1980s, computers started to become more powerful, more flexible in their uses, and more affordable. Traditional mainframe batch processing has been mostly phased out and replaced with distributed, or enterprise, systems made up of multiple desktop computers linked together in local area networks (LANs) which are in turn linked, often over telephone lines, to form wide-area networks (WANs).

Developments in the computer field made possible the creation of digital switches and transmission lines to make telephone communications simpler and cheaper. Introduction of the first digital trunk lines in the 1960s and the first digital switches in the 1970s marked the beginning of the transformation from analog to digital telephony. The goal of this transformation was the creation of an international integrated digital network (IDN) that could provide end-to-end digital connectivity.

The overhead and time required to transform digital-stream computer signals to analog signals for transmission over the telephone network causes sluggish, sometimes unacceptable performance on computer networks. The need to transmit a variety of data types, including voice, video, music, print, and numeric, continues to drive a requirement for more flexible communications through the phone network. The demand for faster and more accurate transmission of computer information over telephone systems has pushed the move to a completely digital phone network, the Integrated Services Digital Network (ISDN).

By definition, ISDN calls for the transmission of video, data, and voice over the telephone network. But seamless digital connectivity is only one benefit of ISDN. A variety of services are available on the ISDN, and they can be accessed at will by their potential users.

Seeing the need for development of ISDN, standards bodies around the world, led by the CCITT, have set key standards for ISDN implementations. Standards deal with the description, concept, and architecture of ISDN and provide for the ultimate evolution from analog telephony and digital distributed computing to one network standard, ISDN.

Proposed implementations of ISDN define standard devices and interfaces, called reference points, between the devices on the ISDN. Four major types of ISDN channel are defined in the standards, as are the BRI and PRI access interfaces. In the chapters that follow, you'll learn all about these types of channels, and the access interfaces that provision them.

chapter 2

ISDN Structures and Functions

In Chapter One, we introduced most of the concepts that make up the ISDN without going into much detail on how all the pieces fit together. In this chapter, we'll take a closer look at how information gets passed over an ISDN, and how users can access the functionality of the ISDN.

Transmission Structures

The ISDN user contracts with an ISDN access provider for a digital pipeline to allow him or her to access the ISDN. The size of this pipeline is predetermined, and the user can consume any or all of the available bandwidth as needed. To meet these needs, ISDN access providers supply *access interfaces* of two basic types. These interfaces comprise combinations of *logical channels*.

ISDN Channels

In the traditional telephone network the user is connected to the network central office (CO) by a local, or subscriber, loop. A local loop consists of one analog channel that is used for signaling to the network—dialing a call, for instance—and for information transfer, which may be a conversation, sound, video, or binary data.

In an ISDN, the local loop carries only digital data, though it may be of any type available in today's environment. The ISDN local loop connects the ISDN equivalent of the CO, the *local exchange* (LE), to the ISDN user's equipment. The ISDN local loop is made up of separate logical channels that are combined to provide the user's interface to the ISDN.

These logical channels are divided into three basic types. Each of these channels exists in a time slot over the local loop pipeline through the process of time division multiplexing (TDM), which we introduced in the first chapter. Furthermore, ISDN channels are categorized by their use, whether for signaling or data transfer, and also by the standard transmission rate for a particular channel type.

The various types of ISDN channels are listed in Table 2-1.

Channel Type	Definition
D-channel	(Device channel) Used for transfer of signaling information between the user and the network and for packet transmission.
B-channel	(Bearer channel) Used for data transmission over the local loop.
H-channel	(Higher-rate channel) Used for services that need higher transmission rates than a single B-channel
B-ISDN	(Broadband ISDN) These channels will enable applications requiring speeds higher than those defined for H-channels. B-ISDN standards allow for transmission rates as high as 622.08 Mbps

Table 2.1 *ISDN channel types.*

D-channel Transmissions

The D-channel is used primarily to carry signals between the ISDN user and the ISDN network itself. Different ISDN user devices, such as telephones, fax machines, and computers, have different ways of connecting with the ISDN, but they all share a common protocol for signaling and receiving signals from the network, and they all use the D-channel for this communication.

Signaling data does not consume the entire bandwidth provided for the D-channel at any given time. Thus, a secondary use for this channel is possible whenever bandwidth is available. This extra bandwidth can be used for packet-switched transmissions such as IP or X.25.

The D-channel may be 16 Kbps or 64 Kbps, depending on the type of access interface provided to the user.

B-channel

The ISDN B-channel is used to carry the information needed for ISDN services. Essentially this means the B-channel is used to carry user information such as digitized voice, video, audio, and binary data. It can be multiplexed to carry any combination of these data types, limited only by the bandwidth of the channel.

The B-channel is the fundamental user data channel described in the standards for ISDN. The B-channel was defined at a rate of 64 Kbps, which was the bandwidth required for effective transmission of a digital voice conversation at the time these standards were set. B-channels are used to make circuit-switched, packet-switched, or semipermanent connections (the ISDN equivalent of a leased line).

H-channel

Some user applications require more bandwidth than a B-channel can provide. Standard configurations of bandwidth known as H-channels provide these higher bit rates. Applications such as video teleconferencing and high-speed data transmission, and the multiplexing of large numbers of lower bit-rate transmissions, all require more than 64 Kbps. H-channel configurations are offered to provide this bandwidth.

An H0-channel is a logical grouping, or its equivalent, of six B-channels operating at a bit rate of 384 Kbps. An H1-channel is made up of all available H0 channels at a single-user interface employing a T1 line. An H11-channel is equivalent to four H0 channels and provides 1.536 Mbps.

Broadband ISDN (B-ISDN)

D-, B-, and H-channels were defined in the first standards for ISDN and provide a considerable improvement in carrier transmission rates. But, things being what they are, even this increase in bandwidth isn't enough. ISDN users want to use the ISDN for video teleconferencing, high-definition television, and other bandwidth-intensive applications that just won't perform acceptably over the original ISDN. These original ISDN standards define what is now known as narrowband ISDN (N-ISDN).

Standards for broadband ISDN (B-ISDN) call for bit rates in the 600 Mbps range. B-ISDN is described in the standards as service-oriented, allowing the transmission of multimedia information integrating data of many types. B-ISDN services are loosely grouped into *communications services*, analogous to traditional phone services, and *conversational services*, which provides ISDN users two-way end-to-end information transfer capability for applications such as video-teleconferencing or high-speed data transfer. B-ISDN transport involves sophisticated *cell-relay* technology, called asynchronous transfer mode (ATM).

Access Interfaces

Access interfaces provide the connections between ISDN users and the ISDN. They are made up of logical groupings of channels provided by the network or service provider. ISDN is designed to allow for multiple information flows over one physical connection, so ISDN access interfaces allow users to switch among available services on demand.

ISDN standards define two different access interfaces, the *basic rate interface* and the *primary rate interface*. These interfaces define the bit rates used for transmission and are differentiated by the number of B-, D-, and H-channels they support.

The Basic Rate Interface (BRI)

The basic rate interface is made up of two B-channels and one D-channel. The designation for the BRI is 2B+D. The BRI D-channel always operates at 16 Kbps. Taken together, the BRI has a total bit rate of 144 Kbps; when additional overhead bits are added, this brings the actual bit rate up to 192 Kbps.

The BRI provides basic phone service to users, and allows simultaneous access to voice communications and a variety of data applications. BRI is intended for small office and home use.

Primary Rate Interface (PRI)

PRIs are provided to users with large capacity requirements, such as offices with digital PBXs or LANs. The PRI in the U.S. is configured to conform to the T1 transmission rate of 1.544 Mbps. The channel structure for the primary rate interface is 23B + D. The D-channel in the PRI has a rate of 64 Kbps.

Needless to say, the PRI is designed for larger companies with multiple ISDN devices in operation. Really big companies using ISDN might even use multiple PRIs, but today only service providers typically consume PRIs in multiples.

USER–NETWORK INTERFACES

ISDN pipelines from the ISDN access provider and the user site have been described as basic or primary rate interfaces. These interfaces are broken down into channels used for signaling the network for services, or for carrying the services provided. That's what the ISDN access provider provides to the user site. What about the ISDN user site itself?

ISDN standards describe the user–network interface of an ISDN as a combination of *functional groupings*, which describe the different devices on the user premises that use the ISDN; and *reference points*, which are logical points of interaction between the functional groups.

Devices on a user's premises may be physical devices such as ISDN phones, or virtual devices that perform an ISDN interface function transparently. These devices are grouped by the role they play in the interface between the customer site and the ISDN access provider.

The ISDN standards define the following devices, or functional groupings:

1. Network termination 1 (NT1)
2. Network termination 2 (NT2)
3. Terminal equipment type 1 (TE1)
4. Terminal equipment type 2 (TE2)
5. Terminal adapter (TA)

Each one of these types of devices performs a set of required functions. We'll describe those in detail in this section. But an understanding of the reference points is critical to understanding a diagram of the user-network interface. These reference points are the conceptual points that divide the functional groups.

In other words, reference points describe the interactions between functional groupings, generally keeping associated functions grouped together. The interactions at the reference point are prescribed by protocols, the rules of the road for the transfer of information around an ISDN user site and onto the ISDN. ISDN standards describe various reference points called R, S, T, and U. We'll expand those initials into explanations in this chapter, and show how these functional groupings and reference points can effectively define an ISDN user site.

Functional Devices

ISDN functional devices fall into the following categories

- *Network termination type 1* (NT1): Represents the physical termination of the ISDN user's interface. NT1 may be provided by the ISDN access provider. It handles OSI level 1 functions such as the physical connection between the ISDN and user devices, line maintenance, and performance monitoring. The NT1 supports multiple channels in the BRI and PRI and handles the multiplexing of bit streams using time division multiplexing (TDM). An NT1 is the ISDN equivalent of the modular plug in the old analog phone system, where multiple ISDN user devices can be connected to the ISDN network.
- *Network termination type 2* (NT2): A device which, depending on its level of intelligence, may provide OSI level 1,2, and/or 3 capabilities. NT2s are used for concentration of user ISDN devices or for switching between those devices. Examples of NT2s are digital PBXs, LAN gateways, and any packet-switching devices. NT2s may not be necessary at small sites where ISDN devices can be directly attached to the NT1.
- *ISDN terminal equipment* (TE1): The TE1 is any ISDN end user device which uses ISDN protocols and supports ISDN services. ISDN telephone, ISDN fax, and ISDN workstations are examples of TE1 equipment.
- *Non-ISDN terminal equipment* (TE2): These are end user devices that are not ISDN compatible, such as standard analog telephones.
- *Terminal adapter* (TA): These allow non-ISDN devices (TE2s) to communicate with the ISDN.

Reference Points

Many device-to-device connections are defined in ISDN standards. Each connection type, or interface, requires a defined protocol. Each of these interfaces is described as a *reference point*.

The four most important reference points defined by ISDN standards are the R, S, T, and U reference points. These are defined as follows:

- *R reference*: Describes the interface between non-ISDN terminals (TE2) and terminal adapters (TA).
- *S reference*: Describes the interface between an ISDN terminal (TE1) or terminal adapter (TA) and a network terminating device (NT1 or NT2).
- *T reference*: The interface between a local switching device (NT2) and the local loop termination (NT1).
- *U reference*: Exists between the NT1 and the ISDN local exchange (LE) and defines the standard for communications between the two. CCITT standards define the NT1 device as part of the local network, and therefore do not deal with standards on the local subscriber loop. ANSI standards in the U.S. define transmission standards over the local loop.

ISDN protocols, which we look at in the next section, are the rules that define how legal proceedings can take place at the defined reference points.

ISDN Protocols

In the world of computer technology, and networking in particular, protocols are used to succinctly describe the relations between entities in a communications environment. Protocols are the rules that prescribe how interactions take place. Earlier, we described the different reference points in the ISDN architecture. Each one of these reference points, which operates as an interface, has associated protocols that describe what takes place at that interface.

Planes

As defined by the CCITT, the ISDN architecture is made up of four planes:

1. The *control* or C-plane
2. The *user* plane (U-plane)
3. The *transport* plane (T-plane)
4. The *management* plane (M-plane).

C-plane protocols deal with control communications between users and the network, such as requests for bearer services and establishing and terminating calls. U-plane protocols deal with the transfer of information between user applications. Transport plane protocols handle the physical connections required, while management plane protocols keep an eye on protocol interactions inside and between the planes.

The U- and C-planes delineate the paths used in ISDN for signaling to the network (C-plane) and transfer of user information (U-plane). U-plane protocols, dealing with the B-channel, are basically transparent to the ISDN. The bulk of ISDN protocols deal with the interface between the user site and the network over the D-channel.

OSI and ISDN

ISDN protocols for the D-channel map more or less agreeably to the first three layers of the OSI model.

1. *Layer 1 (physical layer)*: Describes the physical connections between ISDN devices and the network termination device (NT1). Calls for a synchronous, serial, full-duplex connection. The connection may be point-to-point over the BRI and PRI or point-to-multipoint over the BRI.
2. *Layer 2 (data link layer)*: Handles error control over the physical link. Creates the connection between the user site and the ISDN. The ISDN protocol at this layer is Link Access Procedures on the D-channel protocol (LAP-D for short).
3. *Layer 3 (network layer)*: These protocols deal with signaling between users and the ISDN for establishing, maintaining, and disconnecting calls and calls for supplementary services. Layer 3 protocols deal

with user/network signaling over the interface between the user site and the ISDN. SS7 protocols handle signaling within the ISDN.

ISDN CONNECTIONS

ISDN standards require the availability of three types of service for end-to-end connections:

- Circuit switched on the B-channel
- Packet switched on the B-channel
- Circuit switched on the D-channel

The B-channel, which delivers *bearer services*, can be used for circuit-switching, semipermanent circuits, and packet-switching. The D-channel is used primarily for messages between the user and the network related to connection setup and calls for services.

Circuit-Switched Connections

An ISDN voice connection is established over one of the B-channels. Control information necessary to establish the connection is sent over the D-channel. The BRI will allow two voice connections simultaneously over the ISDN, as it has two B-channels to carry bearer services. Even if the two B-channels are being used, the D-channel is still available to carry signaling information for additional calls. The D-channel manages calls, or it can be used to invoke supplementary services such as call hold and call waiting, up to 15 levels deep in most ISDN environments.

Packet Switching

For packet switching, a circuit-switched connection is set up on the B-channel between an end user and a packet-switched node using the D-channel LAP-D control protocol. The B-channel may be multiplexed using the time division multiplexing scheme to provide a number of virtual channels. Packet switching capability may be provided by the ISDN itself or by a separate packet-switching network attached to the ISDN. Packet switching may take place over the B-channel, or over the D-channel.

Addressing

The world telephone network depends on a numbering system of addresses for each device (telephone, answering machine, etc.) attached to the phone network. The telephone number you dial to reach someone is the unique address of his or her telephone device. Each subscriber in the phone network must have this unique number to route calls and to let access and service providers know whom to bill, and how much.

In an ISDN, a number and an address are slightly different things. An *ISDN number* is the network number of an ISDN user site. An *ISDN address* is made up of an ISDN number, with some additional address bits that identify a specific device on the user site.

Under this scheme, the ISDN subscriber is assigned an ISDN number that identifies the subscriber site. A digital PBX at this site has this ISDN number as its address. ISDN addresses allow for the direct connections between devices at ISDN sites.

ISDN addresses are made up of a combination of the following:

- *Country code*: A one- to-three digit number which specifies the country or geographic area of the call (country codes are already defined in the existing standards for telephony).
- *National destination code*: Breaks down a country into areas and is used to access destination networks inside the designated country. Comparable to the area codes in use in the U.S. telephone numbering plan.
- *ISDN subscriber number*: The designation of individual ISDN subscriber sites.
- *ISDN subaddress*: Designates specific user site devices on the ISDN.

Basically, each T reference point is assigned an ISDN number, and each S reference point is assigned an ISDN address.

INTERWORKING

With all the standards in place, and with ISDNs operating all over the world, the brave new world of ISDN still hasn't really gelled into one world. Instead we find different nations and continents using slightly different flavors of ISDN. And that's just ISDNs. There are a variety of non-ISDN public networks in operation that predate the implementation of ISDNs.

For the goals of the ISDN to be realized, there has to be a method for communicating among different ISDNs, and between ISDNs and existing network components. *Interworking*, the interaction of different types of networks, is critical to the evolution of a worldwide ISDN. Interworking requires that a set of internetwork translation functions be implemented either in the ISDN or in the network accessed by the ISDN.

Interworking of network addresses between ISDNs and other types of networks depends on the ISDN to route the call to the called party's network, and for the called network to complete the routing of the call to the call party. This approach is called *single-stage* addressing. A *two-stage* addressing scheme is also allowed by ISDN standards but is not recommended. Two-stage addressing calls for use of an interworking unit to handle address number transformation.

To provide a compatible connection between an ISDN and another type of network, certain predefined functions must take place:

- Provide interworking of network numbering plans
- Match physical layer components
- Maintain error and flow control over the interconnection
- Collect billing data
- Provide multiplexing and frame construction conversion

For these functions to be defined, additional reference points and their associated protocols have been outlined in the ISDN standards:

- *K*: Interface with existing non-ISDN network; internetworking performed by the ISDN
- *L*: Same as K, but responsibility for internetworking falls to the non-ISDN network

- *M*: Interface with specialized non-ISDN network requiring adaptation functions performed in the non-ISDN network
- *N*: Interface between two ISDNs; protocols determine requirements for compatibility of service
- *P*: Specialized connection within the ISDN providing access to separate components

At present, these reference points are more targets for future implementation than actual, working entities. But they do point to potential methods for integrating heretofore incompatible ISDNs and pre-ISDN telephone systems.

Summary

Users who hook up to an ISDN must contract with an access provider for a digital pipeline. The size of the pipeline will be determined by mutual agreement between the customer and the provider based on user requirements and budget.

This pipeline is defined as a set of access interfaces. ISDN standards define two access interfaces between customers and ISDN access providers, the basic rate interface (BRI) and the primary rate interface (PRI). The BRI and PRI are defined by the channels that are carried over the interface. A BRI, the basic small-company interface, contains two 64 Kbps B-channels to carry bearer information and services, and one 16 Kbps D-channel used for communication between customer and the ISDN. The PRI is made up of 24 64 Kbps channels, 23 B channels, and one D channel. Certain customers may require a connection made up of multiple PRIs.

The user side of the user network interface is described by a schematic made up of functional groupings, or collections of devices with similar functions, and reference points, which are the defined points of protocol-defined interactions between functional groupings.

ISDN adheres to a protocol architecture based on the concept of planes, which can be mapped generally to the OSI network architecture model. Most ISDN protocols deal with user-to-network and network-to-user signaling over the D-channel. Protocols dealing with the D-channel can be mapped to the first three layers of the OSI network model.

ISDN addressing has been implemented to provide routing of calls. The current ISDN scheme is evolving on the model of the telephone network.

ISDN numbers refer to customer sites, and ISDN addresses refer to ISDN devices used on the customer site. As the ISDN evolves, the issue of ISDN interfaces with preexisting public networks must be dealt with. ISDN interworking with these networks will require some type of address resolution between the separate schemes.

ISDN interworking has been defined by protocols, many of which are still under development. The operation of these protocols is defined by additional reference points which describe the points of interaction between the ISDN and other network types.

chapter 3

ISDN Communications Service Links

The basic concept behind ISDN is end-to-end digital connectivity—that is, a completely digital signal channel all the way from sender to receiver, and vice versa. This idea is relatively simple to understand, but the implementation has been a long time in coming. ISDN standards are based for the most part on existing carrier technology; therefore, to fully understand ISDN's implications, you must understand ISDN evolution in terms of communications services that are already in place. In Chapter 3, we're going to examine various types of existing communications, and their relationships to ISDN.

You've already seen how disparate elements in ISDN have evolved from existing telephone and computer technology. You've also learned that ISDN is a long-planned standard for the union of those two technologies. Here we'll try to explain how this convergence will actually occur.

Any significant technological development is pushed along by factors in the marketplace. The need to lower the costs involved in telephone transmission and switching eventually led to digital carriers and integrated digital switching. Improved performance and falling prices for computers and peripherals have fueled the dramatic spread of desktop and distributed computing. The same general market factors are also driving these related technologies to create a union in ISDN.

Markets respond to many factors, and the interplay of these factors keeps them in flux. Few markets are changing as fast in size and scope as the telecommunications and computer technology markets. In these fields, change has become a constant, and technology consumers are wary. One day's cutting-edge technologies may become obsolete within a period of months. Technologies that represent the perfect solution for one company's needs may have little to offer to others. It takes time, and proven technology, to meet market demands.

ISDN technology has to prove itself with existing technology, as well as integrate new ideas. ISDN has been designed to emulate or incorporate existing technologies, but also to be flexible enough to incubate developing technologies. Here, we'll take a look at the market, and at some of the different ways in which ISDN relates to communications and processing technologies already in common use.

T1 Networks

T1 lines are full-duplex digital circuits designed specifically to carry digital signals. T1 digital carrier lines began to be integrated into the telephone network in the early 1960s. T1 lines were originally installed as trunk lines internal to the telecommunications network, and they were intended to provide increased transmission capability and lower the cost of the telecommunications infrastructure. Digital transmission allows more channels to be multiplexed over a single trunk line than is possible over an analog trunk. As advances were made in digital technology, the high cost of T1 began to erode, and T1 lines were extended to customer sites for dedicated, or leased, line service.

When it was first introduced, the cost of T1 service was very high. Advances in digital technology have brought about rapid improvements in the capabilities of T1, both for telephone carriers and for end users. T1 lines are often used by companies to set up private wide-area networks. On a T1 WAN, the telephone network carrier's central office acts as a hub to switch T1 links that tie together remote stations. Data and voice traffic can be multiplexed for transmission over the links.

T1 Anatomy

The T1 line is made up of 24 64 Kbps channels multiplexed to carry voice and data. The T1 architecture is based on the digital signal hierarchy used in North America to describe telecommunications links. During the setup of the analog POTS, it was discovered that the optimal bandwidth for human voice transmission (that's why they invented telephones) was 56 Kbps. With a few overhead bits for transmission control, the optimal channel required was 64 Kbps. Digital transmission cuts down on the overhead required during transmission and allows use of the entire 64 Kbps. The 64 Kbps is the basic unit of the digital signal hierarchy, the DS-0.

- DS-0—64 Kbps
- DS-1—1.544 Kbps

T1 Hardware Requirements

T1 links, at the physical layer, may be of diverse types. A T1 connection could be made over coaxial cable, fiber optic cable, infrared, microwave, satellite, or, most commonly, over an enhanced metal local loop.

To enhance the local loop, the telephone company places units at appropriate intervals along the customer loop to regenerate the digital signal. Thus, the installation of a T1 link does not usually require the installation of upgraded cabling, but does require additional hardware to upgrade the loop.

For connecting a T1 line to customer site hardware, the following equipment is required:

- Channel service unit (CSU)—This device represents the actual interface between the T1 link and the customer site. The CSU maintains line quality, monitors the actual connections across the user network interface, and acts as the physical termination point for the T1 line.

- Data service unit (DSU)—The DSU is responsible for the actual conversion of LAN and voice signals into the digital signals used by the T1. The DSU is connected on one side to the CSU, and on the other to *customer premises equipment* (CPE) such as LAN bridges and routers and multiplexing devices.

The CSU and the DSU are commonly combined into a single device with connections for the T1 on the CSU side, and LAN connections on the DSU side. This device is the physical manifestation of the user network interface between a customer site and the local telephone carrier.

- Multiplexers—Devices responsible for the channeling of circuits in the T1. A multiplexer allows a mix of voice and data over the same link.
- Bridges and routers—These well-known LAN devices usually connect LAN traffic to the DSU.

T1 Costs

Most telephone carriers today offer T1 services to their customers, where T1 circuits will typically be leased on a month-to-month basis. An initial setup charge is standard, and the actual month-to-month charge for T1 links is determined by the distance of the link. Charges are usually based on a per mile rate, plus a monthly service charge. T1 lines are used primarily for links within one telephone Local Access and Transport Area (LATA). If the T1 link spans multiple LATAs, then charges are based on the services of the linked LATAs, and may include charges from an interexchange carrier (IEX) as well.

Prices for T1 local service average about $2800 per month nationwide. Long-distance T1 usage drives rates up rather sharply. The cost of a coast-to-coast T1 link may be as high as $25,000 per month.

T1 and ISDN

The ISDN PRI was deliberately described in CCITT standards to conform to the T1 architecture. ISDN, of course, provides much more flexibility than T1 service. Also, ISDN tariffs are substantially lower than those for T1 service.

T1 multiplexing allows for the sharing of single channels for voice and data, as does ISDN. ISDN, however, allows the user to configure combinations of 64 Kbps channels for higher-bandwidth requirements.

PRIVATE BRANCH EXCHANGES (PBXS)

Private branch exchanges, in the communications world, are the user site equivalent of the central office. A PBX is a telephone exchange at an office or business that provides circuit switching between the various extensions and devices, and provides an interface between the public communications network and the local site. PBX devices can be used to connect devices for data transmission, to connect voice calls, and to provide access to public network services such as frame relay and X.25 packet service.

PBX Connections

Devices

As you might imagine, almost any communications or data processing device can be connected to a PBX. Incoming voice transmission may be switched by the PBX to analog phone sets or digital telephones. PCs and workstations may be separately connected to the PBX, or they may use the same wire pair as the analog or digital phone set. PBXs have long been used to link mainframe devices in an SNA-type mainframe network environment.

Control

A PBX may be manually operated, as is done at most small businesses, or it may be completely automated. Management and control functions are built in to the PBX so that an attendant or operator can provide maintenance and error-checking functions as well as switching capability.

PBX Networks

A PBX may operate by itself in a corporate environment, but it is most probably linked to other PBXs to provide a distributed wide-area private switching system. PBXs on a small exchange may be linked to higher-level PBXs that link multiple exchanges. Through a hierarchical arrangement of multiple PBXs operating at different levels, a private distributed PBX network may be established. Multiple PBXs are usually connected over leased or dedicated lines. The combination of leased lines and multiple PBXs provides organizations with the equivalent of a private telephone carrier network.

Centrex

Most public telephone carriers offer what is known as Centrex service. With this type of service, user devices such as telephones, fax machines, mainframes, and PCs are all directly linked over customer access loops to the public phone network central office (CO). Users could use a private PBX network or a Centrex system and get equal functionality. The major difference in the two systems, as it appears to the end user, would be the access number (called an escape digit) required to leave the PBX network and access the public carrier lines. A PBX network requires the escape digit for outside access. The user of a Centrex system, using carrier-provided switching, has transparent access to the public carrier network.

PBX and ISDN

PBXs are used in an ISDN environment in much the same way that they are used in the analog telecommunications environment.

ISDN PBX

An ISDN PBX may perform the same function of an analog PBX, but the functions are described differently. User devices attach to the ISDN PBX over a two wire-pair connection at the S interface. This connection supplies each device with a BRI connection, or some segment of a BRI. The ISDN PBX is connected to the exchange carrier switching office (CO) over the U interface. An ISDN PRI provides the communications channel between the user site and the CO. The PBX will provide network termination (NT1) and customer premises switching (NT2) functionality.

Centrex and ISDN

In an ISDN Centrex system, BRI connections are made from user equipment directly to the local CO. A user device would be connected over the S interface to an NT1 at the user site and directly over the U reference point (local loop) to the Centrex carrier. The channel between user and CO would be a BRI throughout, and the Centrex network would supply the digital switch necessary for connectivity.

Local Area Networks (LANs)

Considerable work has been done to design and deploy devices to provide interconnections between LANs and ISDN. ISDN is by its nature a wide-area network service (WAN). Standards encourage the linking of LANs to ISDN to provide wide-area network service. ISDN can, if desired, provide LAN services to a user site if parameters for local network performance allow.

LANs and PBXs

Traditionally, an organization has computers interconnected over a local area network (LAN). Telephones are connected to a local PBX. The two networks, voice and data, are not connected. The PBX is connected to user devices (phones) at the user site, and connected to the local exchange carrier (LE) over the local subscriber loop. Computers are connected by cabling specified physical links into LANs. LANs can be connected by bridges and routers, devices which can direct traffic from one LAN into another.

In this type of configuration, the LAN and the PBX do not normally interact. However, a gateway computer can be configured to make remote links in a WAN through the PBX. Given recent developments in computer telephony, however (and Novell's Telephony Services API or TSAPI), the gap between LANs and PBXs should narrow dramatically in the next two or three years.

ISDN LANs

An ISDN LAN would operate in much the same way as the Centrex PBX. In fact, Centrex data services are offered by most exchange carriers. ISDN services can be supplied to networked PCs connected to ISDN over the BRI or PRI. Individual PCs in this configuration are linked to the ISDN provider either directly or through a PBX in the manner just discussed in the section on PBXs.

Performance is a major issue in this type of configuration. Ethernet LANs offer transfers at 10 Mbps, and 100 Mbps LANs are not uncommon. An ISDN LAN operating over B- or D-channels could not exceed the B-channel's 64 Kbps.

Nevertheless, this configuration could be useful to organizations just migrating to LAN technology, and is a good first step to total integration of ISDN services.

Centrex LANs

Centrex data service, offered by most local exchange carriers, is a popular transmission mechanism which integrates voice networks and data networks. A Centrex LAN functions in much the same way as a Centrex PBX. Computers and devices are linked directly to the CO, which acts as the local bus for the LAN. Voice devices (telephones) also directly linked to the CO. Typically, one telephone and one computer share a single local loop.

Centrex LANs depend on a device called an *integrated voice/data multiplexer* (IVDM). On the user side, the IVDM changes voice and data traffic into an integrated digital stream. On the network carrier side, another IVDM decodes the digital transmission into voice and data transmission, which is switched accordingly to either a voice or data switch for delivery to the destination on the LAN.

LAN Interconnection

An ISDN is by definition a wide-area network (WAN). The most LAN-related use for ISDN is to provide a link between remote users and LANs, or to provide a link between remote LANs.

Integrated Voice/Data LAN (IVDLAN)

The IEEE 802.9 committee has presented standards for the integration of voice and data traffic on LANs that has broad implications to ISDN. IVDLAN standards define a connection between user devices (IVDTE) and an ISDN. 802.9 standards call for IVDTEs linked to an access unit (AU) which is in turn linked to the ISDN.

Services provided by the AU travel over channels that correspond to ISDN channels:

- B-channel—a 64 Kbps channel identical to the ISDN B-channel
- C-channel—a circuit-switched channel providing multiple 64 Kbps circuits similar to ISDN H-channels.
- D-channel—a user network signaling channel exactly analogous to the ISDN D-channel
- P-channel—packet channel; provides 802 LAN functionality

TCP/IP (INTERNET)

The TCP/IP protocols are arguably the most widely used protocol suite in existence today. Without going into too much detail, it is understood that TCP/IP protocols deal with network interactions above the data link layer of the OSI model. This means that TCP/IP can run over any data link and physical link protocols. The manner in which TCP/IP is transferred over ISDN is outlined in a series of documents called *requests for comment* (RFCs). RFCs are issued by the Internet Engineering Task Force (IETF), the governing body of the worldwide internetwork called the Internet.

The Internet is a set of LANs and host computers connected by various links such as T1 and T3, satellite, radio, and ISDN. The one thing all these diverse end stations have in common is that they are all using the TCP/IP suite, the standard set of protocols for the Internet.

Internet Protocol (IP)

The fundamental protocol of the TCP/IP suite is IP. IP is a connectionless datagram protocol which is used for communication between network nodes and internetwork gateways, and for communication between gateways. IP is packet oriented, and provides for TCP/IP addressing, delivery, and packet formation. IP is a layer 3 protocol and is supported by the ISDN in much the same way the X.25 packet service, ATM, and frame relay are supported.

Requests for Comment

Provisions have been made for the use of many data transfer methods in the ISDN. TCP/IP protocols are described in documents known as requests for comment (RFCs). RFC 1294 deals with frame relay. RFC 1356 deals with the X.25 public packet switching network. ATM is also described as it relates to TCP/IP transfer.

RFC 1356 describes IP packet transmission on the ISDN. Protocol standards have been adopted by the CCITT and the ISO for interworking between the ISDN and IP networks.

Network Layer Protocol Identifier

An 8-byte identifier, the *network layer protocol identifier* (NLPID), in an ISDN packet identifies the type of higher layer protocols in use. IP has an identifier of 11001100 (0xCC). In an ISDN data packet, the NLPID is the first octet of the call user data field of the packet.

SUMMARY

There is a salient fact about ISDN that has been implied but not yet stated. ISDN represents a wholesale shift in the way public telephone networks are organized, implemented, and operated. Years of evolution in the communications industry have finally provided the building blocks for a nationwide ISDN.

Many of the steps along the way to the adoption of ISDN have offered ISDN-like services, and many more blocks must be put in place to complete a comprehensive ISDN. ISDN represents a fundamental shift in the technology used in the public communications networks. This shift has been taking place more or less gradually. The implementation of digital switching in the local exchange CO was an important step, as was the widespread deployment of T1 digital transmission lines.

ISDN will provide a substitute for many technologies and services now in use. ISDN BRI and PRI are priced in such a way that many dedicated T1 links may no longer be necessary. Customers may receive T1 bandwidth over local loops already in place.

ISDN operations can emulate LAN technology, or it may be used to provide increased LAN services. Connections between LANs and WANs will be much more flexible, and probably cheaper to use, than the ISDN. ISDN can provide for more efficient transfer and increased access to services for users of the worldwide TCP/IP network, the Internet.

Progress toward a nationwide ISDN has been slow and deliberate—too slow for many, too fast for some. One reason for this slow pace is the magnitude of technological and strategic change that must be considered. ISDN marks a convergence of a multitude of related technologies and creates a multitude of considerations. Almost every step taken in the development and implementation of digital and communications technology brings us closer to ISDN.

chapter 4

ISDN TECHNICAL SPECIFICATIONS

ISDN PROTOCOL ARCHITECTURE

ISDN has its own architecture. The ISDN model is segmented into *planes* that describe the structure and function of ISDN. In this chapter we'll examine the four planes that make up this model for ISDN.

We can also map the ISDN model to the OSI model. The planes of the ISDN model cross over the bottom three layers of the OSI model. ISDN is unconcerned with user layers 4–7 of the OSI stack, because it deals solely with network access and not with end-to-end connections between nodes. Generally, applications on the host machines communicating over the network are expected to provide their own end-to-end services (or to draw on protocols, such as TCP/IP, that can provide such services).

ISDN Planes

ISDN standards call for a number of *channels* that, taken together, comprise the ISDN pipeline. The ISDN interface bandwidth is split into combinations of D- and B-channels which carry signaling information and bearer services, respectively. Different sets of protocols are used to describe these different types of channels. The protocol sets make up the *planes*, which are the fundamental units of the ISDN architecture. You'll remember that there are four planes in the ISDN model, the *control* plane (C-plane), the *transport* plane (T-plane), the *user* plane (U-plane), and the *management* plane (M-plane).

The D- and B-channels represent the paths for signaling and user information through the ISDN pipeline. These channels have distinct purposes, uses, and protocols which define them. Since B-channels carry information or bearer services for users, the B-channel is described in the ISDN User plane. Physical standards are defined by the CCITT for ISDN. The Transport plane protocols describe the nature of physical connectivity. Protocols of the management plane have been described as the traffic control protocols. The management plane makes sure that ISDN traffic is directed to, and handled by, the correct plane.

OSI and ISDN

ISDN is used for user-to-user communications, and for user-to-network communications. A working knowledge of the planes of ISDN and their relationship is helpful in describing these functions. For describing intercommunication between user networks already in place, however, we must compare the planes of the ISDN architecture to the OSI model.

Although ISDN planes contain the functionality of the layers of the OSI model, there are certain aspects of ISDN that cannot be fully explained in terms of the OSI model. The most important of these aspects deal with the relations between ISDN protocols, the setting up of multimedia calls over ISDN, and the setting up of conference, or multipoint, calls over ISDN.

A look at Figure 4.2 shows another important aspect of ISDN when compared to the OSI model. ISDN is concerned strictly with network operation, and as such is described completely at the bottom three layers of the OSI stack. Layers 4–7 of the OSI stack deal with connection management and end-to-end connectivity. ISDN expects higher-level functions to be provided by the hosts involved in the communication.

We can also see that different protocol sets are necessary to define the B-channel and the D-channel above the physical layer. Both the D- and B-channels use the same interface at the physical layer, so the same standards and protocols apply. Above the physical layer, however, different protocols apply to these separate channel types. In fact, most CCITT protocols for ISDN deal with user signaling over the D-channel.

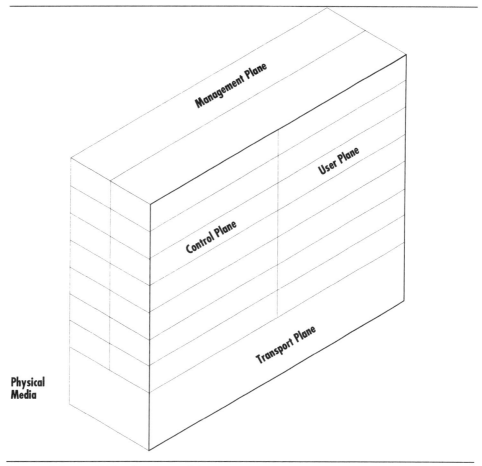

Figure 4.1   *ISDN architecture model.*

Physical Layer Protocols

Layer one of the OSI model concerns itself with the physical characteristics of a network connection. The ISDN physical layer, corresponding to OSI's layer one, has the following functions:

- Encoding of digital data
- Duplex transmission over the B-channel
- Duplex transmission over the D-channel

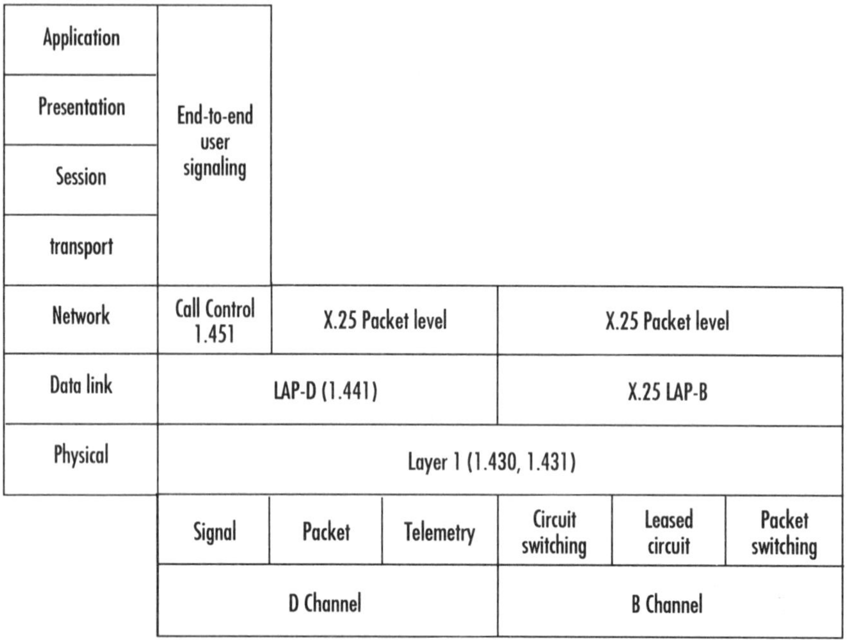

Application	End-to-end user signaling					
Presentation						
Session						
transport						
Network	Call Control 1.451	X.25 Packet level		X.25 Packet level		
Data link	LAP-D (1.441)			X.25 LAP-B		
Physical	Layer 1 (1.430, 1.431)					
	Signal	Packet	Telemetry	Circuit switching	Leased circuit	Packet switching
	D Channel			B Channel		

Figure 4.2 *OSI and ISDN.*

- Multiplexing of BRI or PRI connections
- Activation and deactivation of the virtual circuit
- Provision of power from NT1 to terminal
- Terminal identification
- Faulty terminal isolation
- D-channel contention/access

Remember that ISDN *devices* are connected at *reference points*. ISDN *protocols* describe the nature of the connection and the interaction that occur at these reference points. Figure 4.3 on page 53 shows the different ISDN devices and associated reference points.˘

CCITT Recommendation I.430 defines the physical layer specifications for the BRI. The BRI, as defined in the standards, supports point-to-point and point-to-multipoint connections. The CCITT has defined physical layer protocols for the S reference point, representing the connection

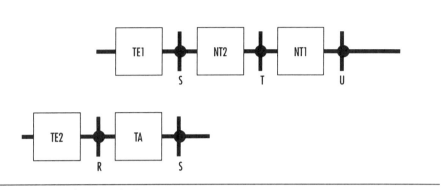

Figure 4.3 *ISDN devices and reference points.*

between TEs, TAs, and a digital routing device (NT2); and for the T reference point between an NT2 and network terminating equipment (NT1).

Configuration

The point-to-point configuration supported by the BRI allows distances of 1 kilometer between the NT device and the connected TE. Multipoint connections are defined as being either short passive bus, or extended passive bus. In the short passive bus configuration, the NT and up to eight TEs can be connected over a single bus. TEs cannot be more than 200 meters from the NT device. The extended passive bus calls for a grouping of multiple TEs separated from one another by no more than 50 meters. The TE grouping can be up to 1 kilometer from the NT device.

B-channels can only be used by one device at a time. User-network signaling ensures that only one TE is assigned to one B-channel at any particular time. Multipoint configurations allowed by the BRI must share the D-channel simultaneously so that messages can be exchanged between the users and the network.

Connections

Standards for the BRI specify full-duplex transmission between the local network and the ISDN provider. The physical connection between the NT and the TE is made over at least two wire pairs. One pair is used for each direction of transmission.

Power

In the traditional telephone system, often called the *plain old telephone system* (POTS), power for telephone devices is supplied by the local phone service provider. In an ISDN, the customer is responsible for powering devices at the customer site. The local loop, the connection between the customer site and the access provider, is defined by the CCITT as a digital transmission facility. This means that no AC power can be carried over the local loop. CCITT standards describe power sources for ISDN devices:

- *Power source 1*: The NT gets power from the network or from a local AC power source, or batteries and supplies power to the attached TEs. In Europe this is sometimes the case; in North America, power is not and will not be supplied by the network provider.
- *Power source 2*: Power is derived locally at the NT device, either from batteries or from a local AC source. Capability to connect all ISDN equipment to central power source or battery backup source. Devices may be connected to the NT and derive power from the NT, or they may be separately connected to a central power supply unit.
- An *ISDN terminal* may plug directly in to an AC or DC power supply.

Digital Signal Transmission

Digital signal transmission is based on the transmission of 1 and 0 values. The most common way to transmit these values is to use different voltage levels on the line to represent one or the other of the binary digits. An encoding method is implemented which will designate a digit and its corresponding voltage. The encoding method for the BRI is called *pseudoternary coding*. Pseudoternary coding dictates that a 0 is always represented by either a positive or negative voltage, and a 1 is always represented by no voltage. The binary 0 pulses must alternate in voltage from positive to negative. Certain bits in the BRI frame are used to balance the voltage on the line, ensuring that there is no net DC component on the line.

Frames

The BRI is a synchronous time-division multiplexed structure. This means that transmission across the physical media takes place within sections of bits called *frames*. Each BRI frame contains 48 bits. For a BRI in the configuration of 2B+D, the total bit rate is 192 Kbps, allowing for transmission of 4000 frames per second. Each frame carries 16 bits for each B-channel, and 4 bits for the D-channel. The bits are interleaved in a particular order in each frame as described in the following:

Channel:	B1	D	B2	D	B1	D	B2	D
# of Bits	8	1	8	1	8	1	8	1

The BRI can be configured as 1B+D, or possibly as only a single D-channel. If one of these optional configurations is used, then the unused bits in the frame for that channel are filled with 1's so that no signal is transmitted.

The other 12 bits in the frame deal with handling and timing of the frame:

- *E bits*—Frames traveling from the NT to the TE carry E-bits which repeat the bits last transmitted on the D-channel. E bits manage contention for the NT by the attached TEs. Since only one device may use a B-channel at any one time, there is no problem with contention on the B-channel. But all devices must share the D-channel for signaling. TEs monitor E bits to know if they can keep transmitting. If a transmitting TE receives an E bit with a different value from its last D bit, it knows it no longer controls the channel and has to stop transmitting.
- *L bits*—DC balance bits enforce the requirement for an even number of 1 bits in the frame, thereby making sure that there is no net DC current. The L bit will be 0 if preceded by an odd number of 0's and 1 if preceded by an odd number of 1's.
- *F bit*—A 0 bit used for the beginning of a frame. Each F bit is followed by an L bit to balance the voltage on the line. The F bit/L bit configuration acts to signal the beginning of the frame to the receiver.
- *A bit*—Activation bit. Used to activate or deactivate a TE device.
- *FA bit*—Auxiliary framing bit. Always set to zero unless used in multiframing.
- *N bit*—Reserved for future use in multiframing. Always set to 1.
- *S bit*—Reserved.

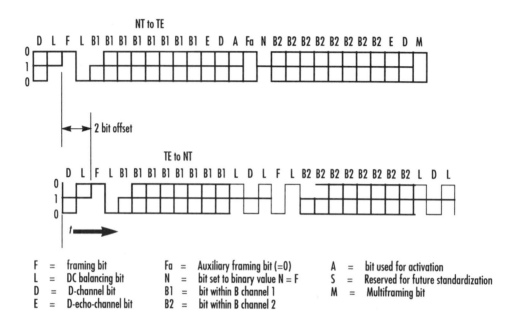

Figure 4.4   A BRI frame.

Contention

As mentioned earlier, E bits help to manage contention between TEs for the S or T interface. B-channels are always assigned to one TE or another, so there is no need for contention resolution. Contention resolution on the D-channel of the BRI is handled as follows:

- A TE with nothing to transmit sends out a series of binary 1s. In the BRI encoding scheme this means no signal on the line.
- The NT echoes back the signal as an E bit with a value of 1.
- A TE wishing to transmit monitors E bits. If it hears enough E bits with a value of 1, it assumes no transmission on the line and transmits.
- If a TE detects E bits with different values than it is transmitting, then it assumes another TE is transmitting and breaks off contention for the D-channel.

The U Interface

CCITT standards do not address the U interface, the connection between the NT1 and the local exchange carrier (LE). These standards have been described in the U.S. by the ANSI standard T1.601. These standards call for serial, synchronous, full-duplex point-to-point connections at the U reference point. ANSI standards allow for the use of existing twisted-pair local loop wiring in the BRI.

The Primary Rate Interface (PRI)

The PRI physical layer is defined in CCITT Recommendation I.431. These standards call for full duplex, serial, and synchronous transmission using two wire pairs. Unlike the BRI, the PRI standard supports only point-to-point connections. The PRI is usually defined at the T reference point. At the T reference point, a digital PBX or LAN concentration device controls multiple ISDN terminals and provides them multiplexed access to the ISDN. The PRI is based on the DS1 transmission structure used in North America for T1 transmission services.

The PRI multiplexes 24 64 Kbps channels in a normal configuration of 23 B+D, where the D-channel is a 64 Kbps channel. Some configurations are made up of 24 B-channels, where D-channel services are provided over an additional PRI.

Framing

PRI frames are made up of one F bit plus one byte of information from each of the 24 channels of the PRI. This makes up for a total frame length of 193 bits per frame. A transmission rate of 8000 frames per second adds up to a total bit rate of 1.544 Mbps.

Framing bits are organized into *multiframes* that provide information about synchronization and error checking. The 24 framing bits provide information about synchronization, frame checking, and maintenance.

Six of the bits in a multiframe form a repeating pattern. If a receiver loses synchronization of frames in a transmission, it just needs to identify this pattern in five consecutive multiframes. This six-bit pattern is called the *frame alignment sequence* (FAS).

Another six-bit sequence is the remainder from the cyclical redundancy check and is used to determine errors at a bit level in the previous multiframe. This six-bit sequence is the *frame check sequence*.

Twelve of the multiframe's framing bits are used to form a side channel used for network management and for messaging. This 4 Kbps channel is called the maintenance channel.

Timing

Density requirements for the PRI specify that no more than fifteen 0 bits occur consecutively on a link, and that at least one bit out of every eight bits is set to a 1. The coding method recommended to provide the proper number of 1 bits is called bipolar 8 zero substitution (B8ZS).

When using B8ZS, if an all-0 set of 8 bytes occurs, it is replaced by the bit pattern 00011011. B8ZS encoding represents a binary 0 by no line signal, and a binary 1 represents a positive or a negative pulse. Binary 1 pulses must alternate in polarity. B8ZS encoding ensures that the wire maintains neutral polarity.

H-channels

The PRI may also be configured to allow H0 or H1 channels as well as B-channels. The PRI can support three H0 channels with a D-channel, or four H0 channels with no D-channel. An H0 channel is the equivalent of six B-channels and can be made up of any six B-channels available on the PRI. The H11 channel uses all 24 channels on the PRI.

The U Interface

For the PRI, there is no discernible difference between the T reference point and the U reference point. Standards across the U interface call for two wire pairs, a traditional T1 link, and standards across the S and T reference points carry the same requirement. Both carriers provide full duplex transmission over two wire pairs.

The D-Channel Data Link Protocol

The data link layer of the OSI model is where reliable communications between physically connected machines takes place. Protocols at the data link layer deal with the setup, maintenance, and disconnection of this type of communication. In an ISDN, all of this activity takes place over the D-channel. For this reason, ISDN data link protocols deal almost exclusively with the D-channel.

ISDN's data link protocol is the *link access protocol-D* (LAPD). LAPD is defined in CCITT standards I.440 and I.441. LAPD is a bit-oriented protocol based on the *high-level data link control protocol* (HDLC) defined in OSI standards. Like all bit-level protocols, LAPD transmits as a stream of bits defined in a structure known as a frame. At the receiving end of a transmission, information in the frame is interpreted bit by bit as it comes off the transport medium.

LAPD's purpose is to prepare and transmit information between ISDN layer 3 entities. LAPD uses the D-channel to define logical connections between users (TEs) and the network across the S reference point and between users and the network across the T reference point.

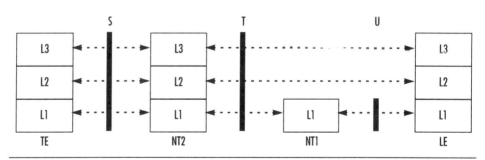

Figure 4.5   *LAPD connections.*

LAPD Services

LAPD is designed to provide two types of service. LAPD must be able to handle multiple terminals on the user-network side of the S or T reference point, and it must be able to support communications between multiple layer 3 protocols operating on the ISDN. Two types of service are provided:

- Unacknowledged information transfer service
- Acknowledged information transfer service

Unacknowledged Transfer

Transfers frames with no acknowledgment. Connectionless datagram service does not guarantee delivery and provides no monitoring, flow control, or error control. Supports either a point-to-point or a broadcast transmission. With almost no overhead, unacknowledged transfer is quick and dirty.

Acknowledged Transfer

Creates a logical connection between LAPD layers. Like all acknowledged transfer protocols, LAPD transmission involves a setup or connection establishment phase, a data transfer phase, and a connection termination phase. In the setup phase, two ISDN entities agree to exchange information. A connection request from one entity is acknowledged affirmatively and a logical connection between the entities is established. During the data transfer phase, the data is monitored for sequencing, error control, and flow control.

These types of operation may occur simultaneously on the D-channel. Multiple acknowledged transmissions can be set up at the same time, as acknowledged operation supports multiple logical LAPD connections.

LAPD Frame Structure

Like any data link, bit oriented protocol, LAPD transmits information in frames. LAPD frames have the following fields:

- Flag—Indicates the beginning and the end of a frame. Always a set of 8 bits in the following pattern (01111110).
- Address—Important to function of D-channel. LAPD addressing allows for the multiplexing of the physical connection between ISDN devices and the network.
- Control—Specifies the type of frame transmitted
- Information—Carries control signaling and packet information.
- Frame check sequence—Used to detect errors in transmission with cyclical redundancy checking.

Flag	Address	Control	Information	Frame Check Sequence	Flag
8 bits	16 bits	8–16 bits	Variable up to 2080 bits	16 bits	8 bits

Figure 4.6 *An LAPD frame (fields and lengths).*

Flag

Flags indicate the beginning and the end of an LAPD frame. The flag at the end of a frame can also be the indicator of the beginning of the next frame. The LAPD flag is 01111110.

One common problem with bit-level flags is that this sequence could occur naturally as a combination of consecutive bit values in other fields of the frame. To avoid this problem, LAPD uses a process known as *bit stuffing*, or zero-bit insertion and removal. Using this process, LAPD counts the 1 bits it transmits. If five 1 bits occur in a row, then LAPD inserts a 0 bit, insuring that the flag sequence only occurs where it is needed.

Address

The structure of the LAPD address field is where LAPD gets its ability to multiplex several logical connections over the physical layer. An LAPD address is formally called a Data Link Connection Identifier (DLCI).

LAPD handles two types of multiplexing. LAPD allows multiple devices at the customer site simultaneous access to the physical link, and it deals with both control signaling and packet data transfer. To handle these types of traffic, LAPD uses a two-part address consisting of a *terminal endpoint identifier* (TEI) which identifies devices, and a *service access point identifier* (SAPI), which identifies a layer 3 process operating on a device. The TEI and the SAPI, taken together, make up the DLCI.

TEIs are usually assigned dynamically as a TE device comes online. In some cases they can be assigned manually by the network administrator, but care must be taken to ensure that no two TEs on a customer network have the same TEI.

SAPIs are used to identify a level 3 process in operation on an ISDN device. Four SAPI values are defined in CCITT standards:

- Call control—Manages setup and disconnects on the B-channel
- 16—X.25 packet mode transmission on the D-channel

- 63—Management information
- Frame bearer service transmission on the D-channel

Control

The control field indicates the type of LAPD frame being transmitted. LAPD uses three types of frames:

- Information transfer frames (I-frames)
- Supervisory frames (S-frames)
- Unnumbered frames (U-frames)

I-frames carry user information. S-frames are used to provide flow control and error control data to LAPD; they control the transmission of I frames on the link. U-frames are used to support the unacknowledged transfer capability of LAPD. Logical links are established, maintained, and terminated using U-frames.

Information

Information fields are found only in I-frames. Information field bits may be in any order in the field, but they must, in the aggregate, make up a certain number of full bytes, or octets. The length of the information field may vary significantly depending on the system in use, but may not be more than the 260 octet maximum.

Frame Check Sequence (FCS)

These bits provide some error-checking functionality. CCITT standards for cyclic redundancy checking define the operation of the FCS. With any given block of data r bits in length, there is an associated FCS number of f bits. $r + f$ should, when divided by a predetermined number, have no remainder. If a remainder occurs, then an error condition is detected.

LAPD Operation

As mentioned earlier, LAPD can operate on the D-channel in acknowledged or unacknowledged modes. LAPD, like most bit-level transport methods, is based on transmission of command and response messages.

Acknowledged LAPD Operation

Acknowledged mode requires the transfer of I-frames, U-frames, and S-frames between a TE and the network. Acknowledged mode transfer has three stages:

- Connection establishment
- Data transfer
- Disconnect

Connections

A logical connection in ISDN may be requested by an ISDN device, or by the network by the transmission. Usually this request is in response to a request by an ISDN level 3 process. Peers exchange LAPD connection information, and the connection request is either accepted or rejected.

Transfer

If a connection is requested and confirmed, then information transfer begins. Information is carried across the physical link in I-frames. Error checking and flow control information is also exchanged during this stage using S-frames.

Disconnect

Either LAPD entity involved in a logical connection may terminate the connection. by sending a U-frame containing the disconnect signal. The receiver sends back a U-frame with bits set to accept the disconnect. Both LAPD instances inform the layer 3 entities associated with the connection that it is terminated.

Unacknowledged LAPD Operation

LAPD also supports unacknowledged operation over the D-channel. Unacknowledged operation has no provisions for flow control or error correction. LAPD user information is transmitted in frames called *user information* (UI) frames. An LAPD user passes information to LAPD in a UI frame which is received and passed up to the layer 3 entity. While no acknowledgment is sent, errors are detected and frames in an error condition are discarded.

Additional LAPD Management

LAPD is also responsible for the assignment of TEIs to ISDN devices. Assignment takes place either on startup of the device, or when a request is made for an LAPD connection. LAPD also can negotiate the use of non-default parameters for the transmission. Parameters are defined as standard defaults, but under some conditions LAPD peers can negotiate the use of optional defined parameters.

LAPD Multiplexing

The ISDN BRI supports point-to-multipoint connections. Since multiple ISDN devices may be connected, LAPD must provide services to multiplex any number of logical connections over the D-channel. Additionally, there may be multiple traffic types traveling between devices. LAPD is responsible for addressing the proper device and the correct process on that device.

TE devices are assigned an identifier, already mentioned, called the *terminal endpoint identifier*. Automatic TEI assignment allows users to install and remove equipment from the user network side of the interface without having to manually update database tables throughout the network.

Service access points are the interfaces between level three processes operating on a TE and LAPD. Each of these access points is assigned a unique identifier, the *service access point identifier*.

ISDN standards dictate that network-to-user-interface, or user-interface-to-network transmissions be established between peer services at the *local exchange* (LE) and the *terminal endpoint* (TE). LAPD uses the SAPI to locate and connect the correct layer 3 process or service requested, and uses the

TEI to differentiate between terminals that may supply that process. The combination of the two identifiers gives LAPD a logical identifier that allows the multiplexing of the D-channel to provide these connections. The combination of the SAPI and the TEI is the *data link connection identifier* (DLCI). Unique DLCIs identify the logical communications channels between the user interface and the LE.

LAPD Channel Priority

The multiplexing features that enable point-to-multipoint connections in ISDN can lead to problems in contention. Multiple devices are connected over a single physical interface (BRI or PRI). The physical connection is further multiplexed, as multiple devices may have multiple layer 3 services. The DLCI ensures that the multiple processes may be connected over the user interface in a process we've already discussed.

This multiplexing demands some type of control mechanism to assign priority in transmission between the multiple channels over the single link. ISDN implements a contention resolution scheme called perfect scheduling with prioritization and fairness that is one of the functions of LAPD.

Pseudoternary Coding

We've already covered how LAPD handles device contention on the ISDN interface by its use of pseudoternary digital signaling. This coding scheme prevents devices from transmitting simultaneously based on TE transmission and NT echoing.

Priority Classes

Still, there needs to be a scheme to prevent one TE, or one group of TEs, from dominating the channel. Priority is assigned according to priority classes, and this is a function of LAPD. Since the D-channel is defined as a signaling channel, signaling information is given top priority (Class 1). Non-signal traffic is assigned to class 2. All class 1 frames have an SAPI value of 0. All class 2 frames have a nonzero SAPI.

Within each class, frames are further assigned a normal priority level or a lower priority level. These levels are indicated by a series of contiguous 1 bits that must be interpreted by a TE before it can transmit. After transmitting a frame, a TE moves to a lower priority in its class.

LAPB and LAPD

CCITT standards for the ISDN data link specify that LAPB, the X.25 layer two protocol, may be used for packet-switching transmission on the D-channel. X.25 installations predate ISDN standards, and X.25 is widely used around the world for packet data transmission. Therefore ISDN standards incorporate use of the pre-existing X.25 protocols.

Besides the fact that LAPD has been designed specifically for use on the ISDN D-channel, there are problems rising out of the use of LAPB on the D-channel.

LAPB is used on X.25 networks for setting up a point-to-point connection between a DCE and a DTE. There are no provisions for multiplexing X.25 circuits over the D channel such as are offered by LAPD. LAPB is used to carry X.25 layer 3 information, but it obviously isn't the most elegant method for this transmission over an ISDN interface. Layer 3 X.25 information can be placed in an LAPD frame.

THE D-CHANNEL LAYER 3 PROTOCOL

In the OSI model, functions related to addressing, routing, and delivery of information are contained in layer 3, the network layer. On an ISDN, network functions take place over the D-channel, as this channel is designated for signaling between the user-network interface and the ISDN.

ISDN layer 3 deals with *signaling* procedures established between the user network and the ISDN, *call control*, and access to and control of *supplementary services*. Layer 3 protocol information is carried across the network in LAPD frames.

User-network Signaling

An important fact in regard to user-network signaling is that D-channel signaling between the ISDN user and the ISDN is not the same as the signaling that goes on between entities internal to the ISDN. Internal network signaling is carried out by SS7 protocols, which are discussed later. Layer 3 signaling deals with signals carried from the user network or terminal to the ISDN.

Messaging

Setting up calls on an ISDN, providing call maintenance, and terminating the call are all handled by the exchange of a series of messages between the network and the ISDN user. CCITT standards (I.451) proscribe a common format for Layer 3 communications which is shown in Figure 4.7.

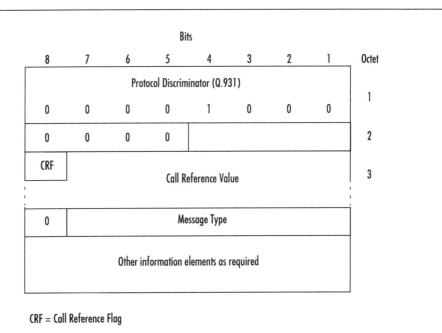

CRF = Call Reference Flag

Figure 4.7   *Layer 3 message format.*

The *protocol discriminator* identifies the protocol the message is intended for. *Call reference* displays a value assigned to a specific active call. There are 33 message types defined in CCITT standards, each with a specific purpose. Each type is assigned a value displayed in the message type field. Various *information elements* follow the message type in an order defined by standards. Table 4.1 on page 68 is a brief list of possible information elements.

Note that list gives only a sample. Information elements may indicate the length of the message being sent, and many additional functions.

bearer capability	facility
called party number	information rate
calling party number	information request
channel identification	packet size
congestion level	progress indicator
date/time	

Table 4.1 *ISDN information elements.*

ISDN Terminals

There are two types of terminals in the user network attached to the ISDN. *Functional terminals* are considered intelligent devices that can exchange messages over the interface between the user network and the ISDN. Stimulus terminals have only minimal signaling capability. A stimulus message represents one particular user event at a terminal. Stimulus terminals may trigger an event in the ISDN, but they cannot invoke it by explicit messaging.

Circuit Mode Calls

One major use of the ISDN is for ISDN telephone service analogous to plain old telephone service. A basic circuit mode call uses an entire B-channel and is set up, maintained, and terminated by an exchange of messages between the user and the ISDN on the D-channel. The control of the call depends on the exchange of a series of messages between the caller and the network, and the network and the called party.

Setup

A circuit mode call connection begins with the sending of a SETUP message from the caller to the network. The SETUP message contains information needed by the network to assign resources to the call and begin the process of connection. The *bearer service attributes* are information elements which tell the network exactly what services are required of the network. Channel identification is also provided in an information element in the SETUP message. The channel identification tells the network the physical interface (i.e., B-channel) the caller wants used to provide the bearer service.

The network will check the contents of a SETUP message and send back a SETUP ACKNOWLEDGE message to the caller. This message will trigger some type of signal to indicate a request for more information, particularly the ISDN number of the called party. The calling party enters the number required, as well as any additional information requested by the network, and forwards all the information back to the network in an INFORMATION message.

Call Setup

When the network receives its additional information, it returns a CALL PROCEEDING message back to the caller, allocates the requested channel, and begins to set up the call to the called party. At the called party end of the network, the network sends a SETUP message to the called party. The called party device's response to the SETUP message will be either a CALL PROCEEDING message to let the network know that the message has been received and call setup is proceeding, or a simple ALERT message indicating that the called terminal is alerting its user to the existence of the call. The ALERT message is analogous to the ringing sound a caller hears on the line when a called party's phone number is dialed.

Connection

When the called party picks up the phone, the device will send a CONNECT message back to the network. This is the signal to the network to complete the circuit over the requested B-channel. At this point the network will return a CONNECT ACKNOWLEDGE message to the called party. The caller's end of the network subsequently returns a CONNECT message to the caller, and the call is connected. The caller and called party can then communicate over the established circuit during a phase ISDN standards refer to as the *call information phase*.

Disconnect

Disconnection is the same as in the POTS. One or both users hang up. This will initiate a DISCONNECT message which is relayed to the other user. The party initiating the DISCONNECT will receive a DISCONNECT RE-

LEASE message from the network, and finally a RELEASE COMPLETE message, which will cause the user terminal to release the allotted B-channel. The other party receives the same series of messages and subsequently releases its end of the B-channel.

Bear in mind that these level 3 messages are used for communication between users and the ISDN. Messaging internal to the network is being handled during all phases by the SS7 signaling system used between entities inside the ISDN.

Packet Mode Connections

Three types of packet mode connections may be established by the layer 3 protocols:

- Circuit mode call to a remote packet handler
- Access to an ISDN packet handler over a B-channel
- Access to an ISDN packet handler over a D-channel

In order for data to be transferred in packet mode, there must be layer 3 protocol support during the data transfer phase of the call. Packet-switching nodes must have access to a service which can set up the virtual circuit needed for the connection.

Circuit-mode Packet Calls

Circuit-mode packet-switched connections are established over the B-channel in much the same way as a circuit-mode call. The caller is really using ISDN services in order to gain access to a packet network. The called party will be an access port on the packet network, called an *access unit* (AU). The AU in this case has the duties of the calling and the called party. If an ISDN terminal makes a call to the AU, the AU initiates the setup of the packet relay. If a packet node makes a call to an ISDN terminal, the AU sets up the call.

ISDN Virtual Circuits

ISDN users may use the B-channel for access to a packet-switching node which itself gives the ISDN user access to the virtual circuit service of the packet network. Recall that the D-channel, as defined in standards, may also be used for virtual circuit service access.

Supplementary Service

Layer 3 messages are also used to *control* and *invoke* supplementary services. Examples of supplementary services would include call waiting, call forwarding, and user identification.

Control

When a user requests supplementary services from the network, messages are exchanged between the network and user as to the service required and the parameters of the service provision. That is, information required to handle the service. A call need not be in progress for control of supplementary services to take place.

Invoke

Invoking a service is a request for dynamic access to supplementary services. Services may be invoked by stimulus terminals or by functional terminals. Where a control message may not be associated with a call in progress, an invoke message always is so related. An invoked service is provided by the ISDN as it is requested.

PACKET- AND FRAME-MODE BEARER SERVICES

Standards for ISDN specify that ISDN may be the carrier for two types of non-voice service already in widespread use. The most prevalent non-voice services in the communications network are X.25 packet mode services and frame relay. X.25 networks have been in operation since the late 1960s. X.25 has considerable built in overhead designed to deal with error identification and correction. At the time X.25 was introduced, this level of error control was required, considering the nature of the physical links involved in telecommunications.

Original standards for ISDN call for the use of X.25 standards for packet-switching of non-voice traffic. Improvements in transmission media and protocols have since created a transmission environment with a very low level of error. ISDN standards released in 1988 recommend the adoption of frame relay as a substitute for X.25 packet-switching. Frame relay provides only a bare minimum of error control in the transmission. The lowering of the overhead allows for a corresponding increase in the setup and speed of frame relay transmission.

X.25 Packet-mode Services

X.25 is a CCITT protocol suite that defines operations between devices in a packet-switching network. X.25 protocols have been used to set up a worldwide public packet-switching network. In a packet-switching network, information is encapsulated in packets that contain addressing, sequencing, and error control information as well as user or application information. The packets are transmitted over virtual channels between X.25 end user devices (DTEs) and packet-switching nodes (DCEs).

X.25 standards were originally released in 1976. By the early 1980s, X.25 was commonly used worldwide for data transfer, particularly between remote terminals and central. ISDN standards call for the support of X.25 networks.

X.25 Protocols

X.25 architecture comprises three layers that correspond to the bottom layers of the OSI model:

- Physical level
- Link level
- Packet level

At the physical layer the protocols specify the physical interface between a DTE and a DCE. X.25 physical layer standards are contained in CCITT X.21, although other physical layer standards may be substituted.

The link level is responsible for the reliable transfer of data across the physical link. X.25's link level protocol is the *link access protocol—balanced* (LAPB). LAPB is responsible for formation of frames which encapsulate data for transmission. The X.25 packet level provides for the formation of data into packets and handles the setup of the virtual circuits over which the packets are transmitted.

Virtual Circuits

X.25 is a connection-oriented protocol. Standards call for packets to be routed over a *virtual circuit* which is established by the level three protocol through the network before packets are transferred.

To set up a virtual circuit, a station makes a call request to the network, asking for a logical connection to another station on the network. All packets that are transferred over this virtual link are identified as belonging to this particular circuit, and they are delivered in sequence-number order.

Two types of virtual circuit are supported by X.25, the *virtual call* and the *permanent virtual circuit*. A virtual call is established as needed by a call setup and call clearing procedure. A permanent virtual circuit is, as named, a permanent virtual connection established by the network. No call setup and call clearing are necessary to use the permanent virtual circuit.

X.25 and ISDN

X.25 has been in common use for almost 20 years. ISDN standards make provisions for the incorporation of X.25 networks into the ISDN. CCITT recommendation X.31 presents the standards by which X.25 and ISDN interoperate.

Essentially, the standards allow for two types of interaction between ISDN and X.25. In one standard, X.31 Case A, ISDN is able to access the services of the X.25 network. In the other, X.31 Case B, the packet-switching capability of X.25 becomes an integral part of the ISDN.

In Case A, the X.25 DTE goes through an ISDN TA to request an ISDN circuit mode connection to the X.25 DCE. The path from the DCE to the destination DTE is established using the level 3 X.25 protocols. Case A procedures cannot be used on the ISDN D-channel, as D-channel signaling terminates at the LE. For this reason, Case A X.25 packet traffic must be carried only on the ISDN B-channel.

In Case B, the X.25 packet-switching capability becomes a part of the ISDN. X.25 DTEs establish the virtual circuit for communication through the ISDN. The ISDN LE either has packet-switching capability or has access to the X.25 DCE. Call setup and control are handled by ISDN. Case B is the standard for ISDN in North America. This standard calls for the transfer of

LAPB frames on the B-channel, and for the encapsulation of LAPB frames in LAPD frames for transmission on the D-channel.

Frame-mode Bearer Services

As the technology has improved, communications links and switching facilities have become much more reliable. The high degree of error control and overhead necessary at the inception of X.25 is no longer required to the extent that it was 20 years ago.

New delivery services, called frame-mode services, have come into common use. Frame-mode services are very similar to X.25, but they eliminate the third layer of the model. Address and connection information is handled by level 2 protocols. The name "frame mode" stems from the fact that level 2 information is transported across the network in frames, not packets.

As a consequence, level 2 protocols are concerned with addressing and multiplexing, in addition to the normal level 2 functions of error control and sequencing.

ASYNCHRONOUS TRANSFER MODE (ATM)

ATM or *asynchronous transfer mode* is a technology based on cell formation and a type of multiplexing that provides users with broadband ISDN services requiring extremely high transmission rates. ATM can supply users with access to this high bandwidth, as it is required, over a B-ISDN channel where voice, data, audio, video, and network signal traffic may be fully integrated.

Standards for ISDN, defining the basic rate interface (BRI) and the primary rate interface (PRI), allow for integrated transmission over the ISDN. The predefined channels of the BRI and PRI provide access to *narrowband* ISDN (N-ISDN) services.

The CCITT, who set standards for ISDN, have designated ATM as the transmission technology to provide transport for *broadband* (B-ISDN) services on the ISDN. B-ISDN standards deal with services requiring very high transmission rates, such as video information services, high-speed data transfer, and real-time video teleconferencing. ATM standards deal with the organization of data into cells for transmission and describe the way bandwidth is multiplexed for carrying B-ISDN services.

ATM technology has been adapted for use over a seamless digital network made up of ATM devices ranging from network interface cards to high-volume digital network switches acting as backbones linking multiple LANs. In this chapter we take a look at broadband ISDN, present the standards for ATM, and look into how the network and communications worlds are coming together over ATM.

ATM Overview

ATM's model is a subset of the protocol model for B-ISDN. Standards for B-ISDN result in a segmented model that maps to the physical and data link layers of the OSI stack. Figure 4.8 on page 76 gives us a two-dimensional look at the model for B-ISDN.

B-ISDN Model

Note that the B-ISDN model is made up of planes just like the ISDN model. The control and user planes cut across the model from the higher layers right down to the physical layer. The management plane is divided in two, with a layer management plane which manages interfaces between each layer in the control and user planes, and a plane management plane which provides a framework for the structured interaction of the protocols in the stack.

ATM Protocol Stack

In this chapter we're primarily interested in the ATM layers related to the B-ISDN planes. Figure 4.8 indicates the layers that describe ATM function—the physical layer, the ATM layer, and the ATM adaptation layers (AALs).

Physical Layer
The physical layer is made up of two sublayers—the *transmission convergence* (TC) sublayer, and the *physical medium* (PM) sublayer. The PM sub-

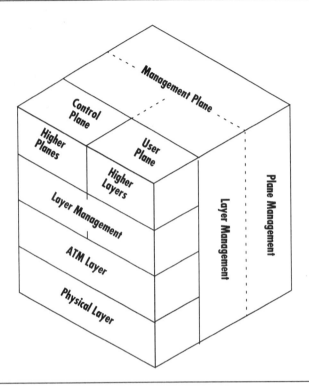

Figure 4.8 *The B-ISDN protocol stack.*

layer interfaces with the physical medium used to provide the link and passes the transmitted bit stream up to the TC sublayer.

The TC sublayer puts ATM cells into the bit stream in frames, recovers the frames, extracts cells from the stream. These cells are either received for the ATM layer, or are handed up to the ATM layer.

ATM Layer

The ATM layer handles multiplexing, switching, and control actions. At this layer the information from the ATM cell header is decoded to provide necessary information for ATM layer actions. Conversely, it is at this layer that the ATM header is created and encoded with transmission information.

At the ATM layer, information is generated and interpreted that provide for the fast switching capabilities of ATM. The ATM layer protocols create the route for transmission using the VPI and VCI fields in the cell header, and direct the stream of cells over virtual channels and virtual paths. The ATM layer passes information up to the AAL layer after stripping the header, and adds the header to cells passed down from the AAL.

The ATM layer generates and attaches the header for each cell passed down to the physical layer, and removes the header on the receiving end.

AAL Layer

This layer is divided into two sublayers, which is appropriate, because this layer has two functions. One of these functions is to provide and interface with higher-level applications and protocols in the B-ISDN stack. The *segmentation and reassembly sublayer* (SAR) takes packets from higher-level applications and segments the information into the right length for insertion into ATM cells. Naturally, the SAR performs the converse action of reassembling high-level messages from fragmented information coming in from ATM cells. The convergence sublayer (CS) presents network services to the higher-level applications and protocols.

Switched Multimegabit Data Service (SMDS)

SMDS is a connectionless switched network service implemented by telephone network carriers. It provides a service that provides customers access over the user network interface to a network carrier's digital transmission facilities. SMDS was originally implemented as a cell-based transmission service connected across the user network interface to T1 or T3 digital links. New implementations of SMDS support network access over traditional 56 Kbps analog phone loops.

The distinction is often made that B-ISDN is a standard, ATM is a technology meant to deliver B-ISDN services, and SMDS is one of B-ISDN's services. CCITT standards define SMDS.

Subscriber-to-network Interface (SNI)

The original standards for SMDS call for transmission of 53-byte cells to the exchange carrier over the *subscriber-to-network interface,* making it the first widespread implementation of cell relay technology designed to run over

ATM. The SNI marks the connection point between the user network and the SMDS carrier. The SMDS Interface Protocol (SIP) is used for communications over the SNI.

Data Exchange Interface (DXI)

Because of the high cost of using legacy packet-switching equipment for SMDS, additional standards have been adopted for the inclusion of packet transmissions over SMDS. Layer 3 protocols in SMDS provide for the transmission of packets of data over the link.

SIP

SMDS is governed by its own set of protocols, the SMDS interface protocol. Layer 3 of SIP deals with the transfer of packets. Layer 2 deals with the creation of cells very similar to ATM cells. The major difference between ATM and SMDS cells is that there is no virtual connection in the connectionless SMDS service, and virtual connection identifiers are not required for transmission. SMDS is a commercial public service, and its connections are predefined. Layer 1 deals with the physical requirements of SMDS.

Signaling System No. 7

So far we've dealt with the kinds of transmissions used for signaling between users and the network, or for transmissions between ISDN devices. For proper ISDN function, there must be a method of signaling and messaging to provide internal control of network activity. SS7 is designed to support digital transmission networks and is specifically designed to operate in an ISDN. Signaling system 7 (SS7) is an out-of-band system used for the exchange of information between entities that make up the ISDN service provider network.

Network Signaling Systems

The establishment and maintenance of calls in a telecommunications network is dependent on some type of signaling system in operation between the different entities and devices that make up the network itself. Information about the routing of calls, information about the status of the connection, and connection control information are all carried by the signaling system. Network signaling provides three major functions to the network:

- Supervision—Monitors the status of the line
- Routing—Provides address information necessary for routing of calls
- Call information—Reports on call status and provides information on the progress of a call

In-band Signaling

Before the telephone system moved to digital switching, all telephone calls were set up over the same channel as the call being made. In-band meant that network signaling and voice communication both took place over the same voice band carried by the user's phone. In-band signaling added considerable time to the call process, as it took from 10 to 15 seconds just to set up a call. Using in-band call setup, circuits between the telecommunications network were allocated one at a time, sequentially. Trunks along the physical route of the call were allocated one at a time as needed. All network resources had to be reserved before a circuit could be established and a call could go through.

All in-band network signals had to be sent over the voice band. Most in-band signaling systems were designed for use with analog voice communications. However, some early T1 lines used a single bit from each frame to carry on/off hook information. Since these bits used for signaling weren't available to the voice band, it took some bits away from the 64 Kbps voice band channel.

Common Channel Signaling

Out-of-band signaling, where network signals travel in a separate signaling channel, is known as *common channel signaling* (CCS). In-band signaling networks have almost all been replaced with common channel signaling systems, such as SS7.

Common channel signaling allows the exchange of information between processor-equipped switches and network facilities and allows for much faster allocation of network resources. Using CCS, the average setup time for a call has dropped from 10-15 seconds to only 3 seconds.

The benefits of common channel signaling are significant. Bandwidth is preserved over the network because no signal traffic shares the actual transmission channels. CCS allows the addition of user services such as 800 service, credit card verification databases, and caller ID services. Perhaps most significantly, CCS brings down the costs involved in calling. In-band signaling requires separate signaling facilities for each circuit established. CCS allows for the multiplexing of network signal information over one channel.

CCS networks make use of two signaling modes. *Associated signaling mode* means that signaling messages follow the same path as the associated call. *Non-associated signaling mode* means that the path for signaling is not necessarily the same physical path as the call.

CCS Networks

CCS networks are made up of standard components:

- Signaling points (SP)—Processor-controlled switching offices
- Signal transfer point (STP)—Switches messages over the CCS network
- Service control point (SCP)—Databases containing information for customer services

STPs not only handle the switching of messages, they provide access to SCPs in the network.

One of the major features of a CCS network is redundancy. STPs are paired throughout the CCS network so that loss of one STP doesn't block calls. SCPs are also paired for redundancy. Several types of links are provided in the CCS network to provide redundant links:

- Access links (A-links) connect SPs or SCPs to each STP of a pair of STPs
- Bridge links (B-links) interconnect STPS in different regions, that is, non-paired STPs
- Cross links (C-links) set up a signal path between paired STPs

All these links are used to provide redundant links through the system and prevent network failure due to a broken signal link.

SS7 Protocol Overview

SS7 is the CCS network designated to be used for ISDN. Like any other system for telecommunications, SS7 is defined by standards and associated protocols that describe an architecture for the system. SS7 protocols are defined in the CCITT Q.700 series recommendations.

Basically, the SS7 architecture is made up of three parts, and their associated protocols.

- Message Transfer Part (MTP)—MTP provides the same service and function as the bottom three layers of the OSI model. MTP is responsible for transfer and delivery of SS7 messages.
- Signaling Connection Control Part (SCCP)—Maps to the network layer of OSI. SCCP provides addressing function and can provide for connection-oriented transfer.
- User and Application Part—Provides for end-to-end signaling for switched voice and data transmission. User and application parts provide function and service analogous to layers 4–7 of the OSI stack.

Message Transfer Part (MTP)

The message transfer part can be described in three levels which correspond to layers 1, 2, and 3 of the OSI stack:

- Level 1 of MTP standards call for full-duplex links of 64 Kbps. 1.544 Mbps rates may also be supported.
- Level 2 of MTP deals with the transmission of frames, called signal units in SS7. Error correction procedures are defined at this level, as are signal alignment procedures.
- Level 3 of MTP provides for message handling in the SS7 network, and for network management. Message handling deals with the routing of SS7 messages through the network. Network management deals with the maintenance of routes through the network. MTP level 3 functions ensure that routes are maintained, link failure and traffic congestion notwithstanding.

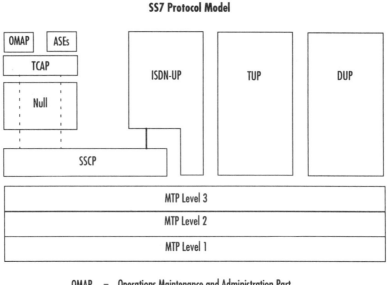

Figure 4.9 *SS7 protocol architecture.*

Signaling Connection Control Part (SCCP)

MTP layers provide connectionless services to the SS7 network. SCCP provides OSI network layer functions not supplied by MTP, such as extended addressing and connection-oriented transfer capability. MTP addressing delivers messages to a node and has limited distribution ability using its own addressing. MTP addressing has limited capability to access services and applications. The growing number of applications and services necessitate the inclusion of SCCP.

SCCP provides four classes of network service:

1. Basic connectionless class (Class 0)—Connectionless datagram service.

2. Sequenced connectionless class (Class 1)—Connectionless datagram service that provides for sequencing of messages.
3. Basic connection-oriented class (Class 2)—Established temporary or permanent signaling connections. Also allows for segmentation and reassembly of messages.
4. Flow control connection-oriented class (Class 3)—Includes all class 2 services with the addition of flow control and provisions for dealing with out-of-sequence and lost messages.

User and Application Parts

User and application parts are independent self-contained entities. Original standards for SS7 provide for the implementation of the ISDN User Part (ISUP), the Telephone User Part (TUP), the Data User Part (DUP), and the Transaction Capabilities Application Part (TCAP):

- The TUP deals with signaling control of telephone communications. The DUP deals with circuit mode data transactions over the network. Neither of these user parts are supported by North American implementations of SS7. ISUP is the SS7 user part concerned with ISDN signaling.
- ISUP uses transport services provided by MTP or SCCP, depending on requirements. ISUP messages are exchanged to provide for the setup and maintenance of an ISDN circuit which will carry user information.
- TCAP provides application-layer services to the SS7 network. TCAP can carry special billing instructions and information, provide customer network control, and support database queries over SS7.

SS7 Services

A topic worthy of its own book might be the different types of SS7 services available. There is an extensive list of SS7 services. They can be loosely grouped into access, or data-based, services and CLASS services.

CLASS

SS7 signaling allows an array of services on the ISDN. The most widespread services allowed by SS7 are called Custom Local Area Signaling Services (CLASS). There are many of these services now in use. In a given area, the services offered depend on the equipment in use and on the network provider's allocation.

The basic CLASS service is called automatic number identification (ANI), also known as caller ID. ANI displays the number of the calling party to the called party during the setup phase of an SS7 directed call. Debate over the implementation of ANI has taken on social and political as well as technical implications. Readers will be familiar with the controversy surrounding implementation of ANI. Some hold that ANI protects users of the phone network from invasions of privacy. Customers may use ANI to set up a blocking procedure on their line, not allowing connections from specific numbers returned by ANI. The opposing view is that ANI takes away privacy from the calling party, which can lead to a number of problems. Luckily, there are less controversial implementations of CLASS services that don't involve the capture and display of the caller's number:

1. Automatic callback--places a return call to the last incoming caller's number
2. Automatic recall—Monitors a busy line until the call is completed and then notifies the caller that the line is open
3. Computer access restriction—Allows access to computer systems only from a predefined list
4. Customer originated trace—Even a customer without ANI could send the number of a harassing caller to the local exchange carrier.
5. Selective services—Allows the customer equipment to be set up to allow calls to be blocked, forwarded, or accepted from predefined lists.

There are many more types of CLASS service. All of these services can be provided through SS7 messaging if SS7 is available to the LE.

Data-based Services

Data-based services deal with provision of services or routing of calls through a distributed network. A list of data-based services includes the following:

- Service—Implementation of SS7 allows for portability of 800 numbers from carrier to carrier and implements faster switching of 800 calls.

- Automatic call distribution—An organization can use automatic call distribution to route certain types of calls through the ISDN to their destination automatically. A call over an 800 number requesting information could be routed anywhere in the ISDN where that information is available.

- Enhanced 911—One major feature of this would be that the caller's ID number and address are displayed to the 911 operator. Additional database services such as availability and location of service equipment and personnel could also be displayed.

- Line Information Data Base (LIDB)—Can provide detailed information about line usage and billing information. LIDB databases could validate credit card charges online.

- Citywide centrex service—Allows PBX services to be provided to business locations by the network provider.

Many other network information databases and services can be accessed using SS7 networks, online telephone directories and online dialing services, for example.

Intelligent Networks

Telecommunications network providers have been working to move from centrally controlled network services to distributed services linked by SS7 access. Distributed signaling networks that can provide customized network services to users require the implementation of more intelligent equipment on the network.

Standards for this implementation are provided by CCITT Recommendation I.312. These standards provide for the establishment of so-called *intelligent networks*.

Intelligent networks allow users of the network to access a variety of network services directly. That's the basic concept of an intelligent network. Users take over control of service access. User control of an intelligent network allows the implementation of more user-specific services on the network.

A logical extension of SS7 signaling is the formation of these distributed service networks. Standards in the U.S. that deal with this extension define what is called the Advanced Intelligent Network (AIN). The goal of the AIN is to allow every end user access to and control of every service on the network.

Presently, the telecommunications networks implement services through switches in the network. Specific services are available only to subscribers who use the associated switch. To justify the cost involved in implementing new switches, network providers must keep services as broad as possible to attract enough users to justify the expense of the upgrade. Intelligent networks will provide a common platform for development of these services, allowing users to develop applications tailored to their specific needs.

As intelligent networks give users and user organizations direct control of network services, it also provides a common platform for the development of these services. Establishment of a common development platform will result in services designed for more specific user functions than are now available.

Broadband B-ISDN

So far, we've been talking about models, standards, and protocols that define what's known as narrowband ISDN (N-ISDN). Many applications, particularly multimedia and video services, need a lot more bandwidth than the standard BRI and PRI. Companies that want these services over ISDN can contract for a user interface that can provide the needed bandwidth. But this can be wasteful activity. By nature, traffic on communications lines is bursty in nature. Companies contract for communications services based on the bandwidth provided. A company that contracts for enough bandwidth to handle peak loads may find that peak loads only occur about 20% of the time. Nobody wants to pay for bandwidth that goes unused.

As part of the ISDN standards, the CCITT has issued standards for broadband, or B-ISDN. Broadband ISDN is said to provide "bandwidth on demand," with the customer paying only for the bandwidth actually used. One key to understanding B-ISDN is to recognize that the B-ISDN interface is not broken out into channels like the BRI and the PRI.

Broadband ISDN services, defined in CCITT standards, are characterized by their need for high bit rates in transmission. B-ISDN is differentiated from the original narrowband ISDN by its need to meet the high bandwidth requirements as they occur.

The BRI offers a considerable degree of functionality to ISDN users, and the PRI multiplies available bandwidth by a significant amount. Standards for BRIs and PRIs define what is now known as narrowband ISDN (N-ISDN). ISDN standards calling for high bandwidth services define a new category of ISDN, broadband ISDN (B-ISDN).

Most B-ISDN services call for much higher transfer rates than the PRI can provide. B-ISDN services such as high-speed data transfer and real-time desktop video teleconferencing need lots of bandwidth, as much as 25 Mbps for compressed high resolution video. A customer requiring these services would need a bundle of PRIs to provide available bandwidth, and transfer of all that information would monopolize a standard N-ISDN channel. Conversely, when high-bandwidth services aren't being accessed, lots of interface bandwidth is going to go unused.

Customers using N-ISDN are paying for available bandwidth. B-ISDN customers have similar requirements for N-ISDN services, but require high levels of bandwidth at some times to allow transfer of services unique to B-ISDN. B-ISDN channels must still handle different tasks such as packet transfer and telephone calls, but standards also call for the integration of bandwidth-intensive services over the same access loop. So the B-ISDN channel must be *scalable*. This means that the channel must allow high bit-rate transfers for B-ISDN services when needed, so that the customer using a B-ISDN channel pays for bandwidth that is used, not predefined bandwidth supplied by N-ISDN channels.

Asynchronous transfer mode (ATM) is the technology defined in the standards to be the bearer of ISDN services. The B-ISDN channel is multiplexed using ATM to provide a single channel over which traditional N-ISDN services and B-ISDN services may be integrated.

Architecture

The architecture for B-ISDN is shown in Figure 4.10. Note that the B-ISDN model is made up of planes just like the ISDN model. The control and user planes cut across the model from the higher layers right down to the physical layer. The management plane is divided in two, with a layer management plane which manages interfaces between each layer in the control and user planes, and a plane management plane which provides a framework for the structured interaction of the protocols in the stack.

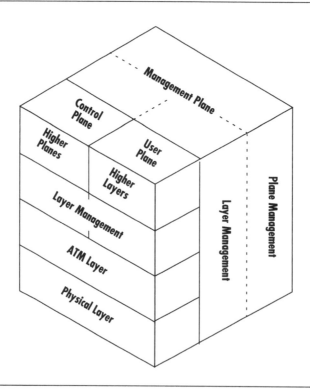

Figure 4.10 B-ISDN model.

Broadband Services and Standards

Some ISDN services will need transmission speeds greater than those that can be delivered by a single PRI. These services are to be delivered by broadband ISDN, or B-ISDN. B-ISDN services are loosely divided into two main groups, communications services and distribution services.

Communications Services

Communications services include all of the services delivered over the traditional analog telephone network, as well as conversational, messaging, and retrieval services:

- *Conversational services* are two-way real time communications between remote user stations. Video teleconferencing is one major service made possible by B-ISDN.
- *Messaging services* enable the transfer of user-to-user mail applications, including voice, video, and document services in any combination.
- *Retrieval services* offer users the ability to access databases of information stored as video, text, images, documents, or data.

Distribution Services

Distribution services will provide users of the B-ISDN access to network services such as high-definition television, pay-per-view television, video libraries, interactive multimedia applications, and news services. They are defined as being *user-controlled* or not. Access to online video databases might be described as user-controlled, because users can choose when they want access and to what they want access at any given time.

Services outside user control would be HDTV programming, which can be understood as analogous to present-day TV programming. The network provides broadcasts of programs and information at scheduled times, and users may choose to access them or not.

Requirements

To implement and utilize the B-ISDN, there are certain criteria, or requirements, that must be met for proper operation. These criteria deal with the bandwidth requirements of B-ISDN services and with the levels of quality in presentation that are required by users. For example, the metal customer loops that connect user sites with the ISDN are supported by N-ISDN standards, but will be unable to carry the bandwidth required for some B-ISDN services. These loops must be replaced, probably by optical fiber loops, in order to enable B-ISDN at the customer site. Limitations imposed by circuit-switching technology also present the requirement for the adoption of ATM cell relay switching to enable many B-ISDN services.

Bandwidth

All of the services of an N-ISDN, plus additional services with very high bandwidth requirements, are to be supplied by the B-ISDN. High-resolution video, a B-ISDN service, requires a channel of approximately 150 Mbps. To support multiple B-ISDN services, an interface must have an aggregate channel rate of 600 Mbps. Optical fiber local loops will support these high bandwidths, giving rise to the possibility that N-ISDN services may continue to be supplied over the metal-based local loop, and that high-bandwidth B-ISDN services will be supplied over an additional fiber loop. Inside the ISDN, all of these bandwidths must be supplied as requested to enable B-ISDN.

SUMMARY

An overview of the technical aspects and requirements associated with the ISDN gives us a fairly comprehensive look at the current state of communications and data processing technology. The mutual dependence of these two technologies has produced advances and developments at a rapid rate.

Standards for ISDN were developed years ago as communications and computer researchers realized a need for and an understanding of the eventuality of the convergence of the two fields. ISDN standards were a response to the perceived need to channel new developments toward a common goal of end-to-end digital connectivity.

The ISDN model is designed to display the multiple channels of an ISDN connection and the function provided by each channel. Through the BRI and the PRI, point-to-point and point-to-multipoint connections are possible.

The BRI and the PRI are offered today by most ISDN access carriers as solutions to business needs. In the creation of their standards, an effort was made to anticipate the requirements for communications and to provide for them.

The BRI, with its 2B+D channel configuration, is the standard interface between the ISDN and a small-business or home user. The BRI provides service that is the equivalent of basic telephone service combined with data service capability, and it can be configured to provide a higher level of service to users than is available over the analog local loop.

The PRI gives a much wider selection of channels and configuration choices. Medium to large businesses and organizations contract for the PRI, or a bundle of PRIs. The much higher aggregate bandwidth of the PRI not only provides an organization more channels for communication, it may also enable high-bandwidth transmissions over combinations of B-channels.

The key to ISDN functionality is in its channel structure. D-channel signaling for setup and control of calls eliminates overhead on the bearer channels (B-channels). Control messages are relayed in LAPD frames. LAPD frames may also contain packetized data, allowing for use of the D-channel as a packet transmission channel.

Public data transfer services offered by telephone service providers are supported by the ISDN. Layer 3 signaling over the D-channel allows the setup of frame mode bearer services. ISDN can provide a link to X.25 packet-switching networks by setting up a communications channel to an X.25 gateway, or by carrying X.25 information in an LAPD frame.

Standards for broadband ISDN (B-ISDN) provide a flexible interface between users and the network. B-ISDN supports applications with very high bandwidth requirements over an interface that is not provided in channels as in the BRI and the PRI. B-ISDN standards call for the use of ATM, a cell relay technology, as the carrier for B-ISDN services.

ATM will provide layer 1 and layer 2 network services over a cell-switched network at very high speeds. ATM is designed as a transmission structure to carry multimedia applications, video, high-quality voice, and high-speed data transfer. A public data transfer service already in place, the switched multimegabit data service, provides layer 3 packet capability and will adapt to run over ATM.

ISDN data link and layer 3 signaling on the D-channel define the standards for user-to-network and for network-to-user messaging. Messages between internal network stations are provided by an out of band signaling system called SS7. SS7 carries user messages across the ISDN. SS7 has its own protocol structure and internal messaging format. It creates an intelligent network which can respond systematically to user requests as well as calls from other internal network devices.

part II

CURRENT PC ISDN HARDWARE AND SOFTWARE

The hardware and software ISDN offerings available for the PC are surprisingly varied and numerous. In this part of the book, we'll examine how to determine your ISDN hardware needs. Then, we'll proceed to examine a number of different kinds of hardware and software offerings that you might use to connect your PC, or a small network of PCs, to an Internet Service Provider (ISP) using ISDN.

The number of options for internal and external interfaces, built-in or external terminal adapter equipment, and other aspects of ISDN hardware can sometimes be confusing. That's why we concentrate on explaining what items of hardware and software are necessary to establish and use a working ISDN connection. That's also why we try to contrast and compare the various options, and to explain the pros and cons of particular implementations or approaches.

In this, the second part of the book, we start you off in Chapter 5 by presenting a questionnaire that can help you to determine your ISDN hardware and software needs. In Chapter 6, we examine the various options available for internal and external ISDN PC adapters, with discussions of the leading vendors and their ISDN interface products. In Chapter 7, we shift our emphasis to examine terminal adapters and network termination equipment. Then, in Chapter 8, we review the various kind of bridges and routers (generally for Ethernet networks) designed to attach small LANs to an ISDN link. In Chapter 9, we change direction to examine the various kinds of ISDN software available, with special emphasis on interoperation with standard network drivers and stacks such as Novell's Open Datalink Interface (ODI) and Microsoft's Network Device Interface Specification (NDIS). Finally, in Chapter 10, we take a quick look at the options available to bring voice and data through your ISDN link, as we examine the offerings for native ISDN phones, and the widgets that will let your POTS phones work over an ISDN link.

Our goal in Part II is to arm you with information about what kinds of ISDN products are available for your PC, to expose you to the various vendors who participate in this marketplace, and to try to share with you our own experiences (and the experiences reported by others) in using this equipment and software. We feel that you'll be much better-equipped to purchase and configure what you really need, if you understand what's available and what kinds of trade-offs you must make in selecting one product over another.

# Chapter 5

DETERMINING YOUR ISDN NEEDS

SO YOU'RE SERIOUS ABOUT ISDN?

If you've skipped Part I of this book and started here, that's OK. The idea here is to briefly explain only what you need to know to decide if ISDN is for you, without causing your eyes to glaze over. If you don't understand an acronym or buzzword, you can check the glossary or return to Part I for the details.

Today, ISDN is one of the "Top 10" buzzwords that's sweeping the computerized world. You may be thinking seriously about an Integrated Services Digital Network for personal or business use, or you may only recently have found out what the acronym stands for. Even now you may still not be sure what it *really* is, how it *really* works, or if it will be *really* cost-effective. Not to worry: You're way ahead of the pack, just by being curious enough to buy this book and learn more about if or how ISDN can benefit you.

Deciding whether to jump into the world of digital telephony (ISDN) from the analog telephony world—or POTS (Plain Old Telephone System), as it's irreverently called by ISDNers—isn't really as complicated as it looks from the outside. Sure, there are a multitude of different ISDN products available and more keep popping up every day. All we can say is "Welcome to the bleeding edge of electronics!" Thankfully, this variety of good products insures enough competition to cover your needs with reasonably priced and well-supported products. As recently as the spring of 1995, this was not the case.

Our advice is: Don't let a plethora of ISDN products overwhelm you. You only need one or two of them to get your connections up and running, and we'll help you choose them.

Thankfully, the telephone companies are fairly well educated in the intricacies of ISDN—or at least, some of their employees are. As one person (who must remain nameless) so cogently stated: "All of the ISDN workers went to the same classes, but not all of them stayed awake and walked out with the same knowledge." At least you'll be able to call your local telephone service provider, ask for ISDN assistance, and eventually find a person who knows what ISDN means. This is generally true if you live in an urban area in the many parts of the United States. Rural areas will have to wait for ISDN service just a bit longer.

How Do You Decide?

The very first step is to answer the following questions truthfully:

Do I really need high-speed, digital phone service to my home or office?

For personal use at home, the only compelling reason to install ISDN is for fast access to the WWW (World Wide Web) and other Internet or on-line services. If you have the desire and can afford the costs, you'll find the speed highly appealing. If you're an on-line worker, you'll find the added costs will quickly be repaid by the savings in connect time and productivity improvements.

For business use at home or in a one-person office, the primary reason to install ISDN is also to obtain speedy access to the WWW and other Internet or on-line services. Secondarily, you'll also get two phone numbers as part of your connections, and will be able to use computerized fax services and digital telephony (answering via computer and digital telephones) just like the big companies do. The cost of ISDN service may be less than or equal to the cost of two business lines to your home or office; this alone may make the switchover worthwhile!

For small businesses, if you have an office with two or more people, each of whom needs fast access to the WWW and other Internet or on-line services, ISDN may be an appropriate way to go for voice, data, and multiple ISDN phones. If your computers are already connected via a network, installing ISDN will be more complicated, but it may also allow sharing your ISDN lines and services using the network.

Small-business owners or operators may find that the costs of equipment and ISDN line installation and monthly fees are offset by increased productivity of your employees. However, if you aren't confident about running your own network without the ISDN, you may find it advisable to hire an ISDN consultant who is familiar with your type of network to suggest the appropriate equipment for your particular needs, and to install and maintain the equipment. Although they can be expensive up front, consultants can save you money by keeping you and your employees working, rather than wasting time fooling around with your network and ISDN problems. Take our word for it--we learned this first-hand!

Can I afford the cost of installation by the phone company?

As of Spring 1995, the cost of installing a single BRI (2B+D) ISDN service varied from $0 to $585 in the U.S. The average installation fee was in the $150 to $250 range. Our own local RBOC, Southwestern Bell, offered a promotional test fee structure that went from no cost for installation if you signed a three-year contract to $585 with no contract, with incremental charges for one- or two-year contracts. By the time you read this, they will have changed these rates again, as will many of the other regional phone companies. Call yours and ask for the current installation charges; ask around and find out if there's any competition in your area, too—you may find an even better deal just by shopping around!

Can I afford the monthly service fees from the phone company?

The RBOCs (Regional Bell Operating Companies) and other ISDN providers seem to be constantly changing their tariffs and therefore their rate structures. Maybe they're not as bad about it as the airlines are about ticket prices, but it's sometimes hard to tell. Rates in the United States currently run from a monthly fee of $25 to $92.50 with no per-call charges, to between $19.50 and $70 per month with per-call usage charges of $0.0042 to $0.15 per B channel per minute.

Can I afford the cost of installation or setup from my Internet Service Provider?

The going rates for ISPs (Internet Service Providers) to set up an individual, non-business ISDN account using SLIP or PPP varies from $25 to $200 across the U.S. Most ISPs charge more for business account setup with multiple user numbers. In their defense, it takes a very knowledgeable network person upwards of 30 minutes per user in setup time. In fact, the process is far from automated at most ISPs. Some of this cost theoretically goes toward the labor involved in the setup process; the remainder is usually intended to help defray the provider's up-front hardware costs.

Can I afford the monthly service fees from my Internet Service Provider?

ISPs are very competitive in their ISDN rates, and therefore frequently change their rate structures. Without going too far out on a limb, we've observed that most ISPs offer two different ISDN services: dedicated and dial-up. Most such connections use either SLIP or PPP protocols, depending on the provider's own server software and its ability to measure and bill for connect time.

Dial-up SLIP/PPP rates generally range from $30 per month with 45 hours of 1 B channel connect time included, with additional hours charged at $1 per; to $29 per month with 30 hours of 1 B channel connect time included, with additional hours at $2 per.

Flat-rate monthly charges for unlimited time—and since nobody can use the connection you're using if you use it all the time, it's often called a dedicated line—range from $150 for uncompressed single B channel to $400 for two compressed, BONDed B channels. The rates for dedicated lines appear to vary—at least in our immediate vicinity—more than those for dial-up lines, so you'll want to shop around as much as you can stand to when looking for such service.

It's also important to remember that ISDN equipment takes less than one second to "connect" to your ISP. Therefore, if you are being billed for "connect" time, you may be able to keep your charges quite low by setting your time-out to a few idle seconds (this is discussed in Part III). In general, when buying ISDN connections, you should expect to pay at least $30 per month and possibly as much as $60 to $100, even if you spend as little as three hours per day on line.

Can I afford the cost of the ISDN equipment for my PC?

For an individual PC, the cost of an NT1 and TA, whether it is an adapter card, bridge, router, or ISDN "modem," will range between $400 and $1500. Some manufacturers have marketing agreements with local telephone service providers or ISPs whereby you can purchase the hardware in a package deal with the service. It is possible to get a combined TA/NT1 for as little as $200 in a package deal.

Generally, you should expect to pay between $500 and $700 for an ISDN adapter card with built-in NT1 and a POTS port. A separate NT1 costs between $200 and $400 with power supply (and sometimes a built-in battery backup). External ISDN bridges or routers generally range between $800 and $1500 depending upon their features.

The bottom line is as follows: An adapter card is about half the cost of a separate NT1 and bridge or router combination, and about two-thirds the cost of a separate NT1 and ISDN "modem." Most of these adapter cards are still new to the scene as of early 1995, so it's reasonable to look for more functionality and ease of use in the future, along with lower prices as economies of scale begin to kick in.

Can I afford the time necessary to deal with my local phone company?

Most RBOCs and other ISDN service providers have ISDN order lines with reasonably knowledgeable personnel to help you with your needs. However, this does not mean that everything will progress smoothly, rapidly, or without your close supervision.

Here's some of what you'll encounter on the way to ISDN nirvana. If your location is near an ISDN switching station, there may be unused phone wire pairs available to you; if not, you may have a very long wait for the necessary hookup. If your location and wiring is less than 10 to 15 years old, and the local phone company isn't swamped with installations when you call, you may get lucky and have to wait only a couple of weeks for installation. Otherwise, it can take up to six weeks or more, even if your location meets most of the conditions above. During this period you will be spending your valuable time on your POTS phone to more than one TPC (The Phone Company) employee trying to get things going or to figure out what, if anything, is happening.

Since ISDN to residential locations is relatively new, there will be more problems than you have with your current analog system. When an ISDN service goes "down," you don't get a weak or buzzy line, you get a big, fat nothing. Zip, zero, nada!

ISDN service is either up and working, or it's stone-cold dead. So, if you lose power to your NT1 device, your ISDN service is dead (that's why battery backup is nice during the occasional power outage). This "dead or alive" characteristic is quite different from POTS, which draws its power from its own sources, not from your house or office wall plug. However, that same plug is exactly where your NT1 or computer gets its power. Remember this when determining whether or not to hire a consultant, or whether or not to go completely ISDN with no POTS phones for backup. Like the mighty mainframe, we think POTS will still be around for a long, long time!

Can I afford the time necessary to install the ISDN equipment myself?

Assuming you are a knowledgeable computer user who feels comfortable installing cards inside your computer, but not an electrical engineer or hot-shot programmer, it can take you from one day to three weeks to get your ISDN equipment installed and running properly.

Using an external NT1 with a serial ISDN "modem" type device, you may get lucky and have it all running in a few hours. If you try to install an ISDN adapter card that is actually a network card in a computer that already has another Ethernet or other network card installed, and you try to get it all running under Windows for Workgroups or NetWare, and the provisioning (setup) on your ISDN line isn't exactly what you ordered, it may take you three weeks or more.

This time frame includes several hours on the phone to the customer support personnel of your chosen ISDN adapter card, to Microsoft or Novell for networking advice, to your telephone service provider for switch and provisioning information, and to your Internet service provider for connection testing and debugging. It also involves seemingly endless changes to your PC's configuration files (e.g., to CONFIG.SYS, AUTOEXEC.BAT, WIN.INI, and SYSTEM.INI), and to its CMOS setup maps.

Warning! This is by no means the worst-case ISDN horror story. Unfortunately, a certain amount of thrashing around happens all too frequently, given the variety of new cards and other hardware in PCs today. This sort of thing happens much less frequently in the Macintosh world, due to the standardization of the operating system and its interface managers for added requirements. In any case, be aware that it's going to take you much longer than you anticipate to get everything up and running smoothly. If you plan for this, and arrange for expert help to be on tap if and when it's needed, you'll lose a lot less sleep during this sometimes excruciating process.

Can I afford to hire a consultant to install the equipment in my location?

If you're getting ISDN for a profitable business that needs your time to keep it profitable, seriously consider hiring a qualified consultant. Do a quick cost/benefit analysis comparing your lost time and lessened efficiency vs. the consultant's fees. Keep in mind that a good consultant can assist you in choosing the proper equipment for your business, both now and when you need to expand.

If you are getting ISDN for personal use only and, like most of us, have more time than money, keep on reading and do it yourself. As the working wife once said to the out-of-work husband, "Honey, your time is valuable but not costly." Choosing the right ISDN equipment and ISP, and knowing how to deal with the phone company, should make your initial ISDN experience less of an ordeal than you may expect.

What Do You Want to Do?

Are you still with us? Good! That means you're either committed to getting your own ISDN service, or you should be committed ... or both. To aid you with your tasks, find the section below that best describes your situation and read it completely and thoroughly. In each of these sections we've condensed the answers to the previous questions and customized them for the specific group it addresses.

Home for Personal Use (ISDN Internet via Computer + POTS Phone)

Here, you already have a computer at home and simply want to get ISDN service so you can access the Internet and other on-line services faster and better. You wouldn't mind the extra telephone line either, since your kids/wife/etc. are always griping about your tying up the line with your modem. You aren't really a hardware type, so installing complicated boards doesn't appeal to you. You already have an Internet account and your ISP offers ISDN PPP dial-up accounts, or you're willing to switch to another ISP for an ISDN account. Your modem is getting old and you want a faster one anyway.

For this scenario, you will need the following:

1. ISDN service from your local telephone company. Installation averages $250. Monthly fee averages around $60.
2. NT1—stand-alone $300
3. External serial port ISDN "modem" $600–1050.
4. PPP software and TCP/IP stack; freeware or shareware.
5. ISDN service from your Internet Service Provider $50 setup plus $30-60 per month.

Bottom line: The total installation and hardware cost is between $1200 and $1650. The monthly fees will run approximately $120 per month.

Home Office/Personal Use (ISDN Voice and Data + POTS Phone)

You are a work-at-home consultant or have a one-person office. You already use a modem with your Internet account and CompuServe, AOL, etc. You have three POTS lines for voice, your computer, and your fax machine. You want to install an ISDN service to replace the data and fax phone lines and to increase your on-line speed and bandwidth.

You feel comfortable installing cards in your PC and are enough of a hacker to alter your CMOS setup without calling AMI. You will either leave your computer on all of the time for your fax machine or you'll not worry about it after hours. You have enough time to spend doing it yourself and want to keep the initial cash outlay to a minimum.

For this scenario you will need the following:

1. ISDN service from your local telephone company. Installation $250. Monthly fee $60.
2. Combined NT1/TA internal adapter card with POTS port $500
3. PPP software and TCP/IP stack and network drivers. Freeware or Shareware.
4. ISDN service from your Internet Service Provider $50 setup plus $30 - $60 per month

Bottom line: The total installation and hardware cost is about $800. The monthly fees will run approximately $120 per month. Eliminating two business lines saves you about $60 per month in phone charges, so you break even on the phone service. Your ISP charges jump from $20 to $60 for a net increase of $40 with ISDN over POTS usage for all lines.

Small Business Office (ISDN Voice and Data, Multiple ISDN Phones)

Your office is relatively high-tech with you and your partner using PCs in a Windows for Workgroups (WFW) Ethernet network. Each of you has a separate POTS line for dialing out. You have separate POTS lines for voice for each of you with no receptionist and no intercom functions. You both want faster access to Internet and other on-line information services which will undoubtedly be offering ISDN connections in the near future. You also want to be able to e-mail more easily to your clients and send/receive large files more quickly. If you grow even a little, you'll need a receptionist and a small PBX phone system. You are adept at working with your hardware and feel comfortable with the amount of your time it will take.

You are at the point where you need to carefully determine if you need an ISDN and/or network consultant or really want to do it yourself. Presuming you want to do it yourself, here is the least expensive and simplest scenario.

For this application you will need the following:

1. ISDN service from your local telephone company. Installation $450. Monthly fee $120.
2. Combined NT1/TA internal adapter card with POTS port, two @ $500
3. ISDN service from your Internet Service Provider, two @ $50 setup plus $60–120 per month.
4. PPP software and TCP/IP stack and network drivers for WFW. Freeware.

Bottom line: The total installation and hardware cost is about $1550. The monthly fees will run approximately $120 per month. If you replaced one POTS line for each person with an additional ISDN number, you saved approximately $60 per month. Your ISP charges probably increased from about $40 to $120 per month. This results in a net increase due to ISDN of

$160. You could further lower this by $30 through using one of the ISDN ports for your fax machine.

You could replace all of your POTS lines with the two ISDN services (four numbers total) if you installed a separate NT1 and used TAs without built-in NT1s in your computers. You could plug your POTS phones into the NT1 as well as your fax. You would need a substantial UPS for your NT1 to keep your phone lines up and going if the power goes out, so the cost would stay about the same for the hardware. The monthly service could be lowered another $30 to $60 depending on how many POTS lines you currently have. This should make the monthly costs for your phone lines less with ISDN, but your ISP charges will still be more, so it isn't a completely break-even situation. However, you are getting the benefit of the ISDN speed and bandwidth.

What Do You Need?

As you have undoubtedly gathered, there are certain elements that are essential for your ISDN service installation to succeed. At a minimum, you will need the following items, information, or services to connect to the Internet and make voice calls via your ISDN telephone service:

1. A properly provisioned (set up) and operational ISDN line from the phone company.
2. A properly configured ISDN account with an Internet Service Provider.
3. Software to link you with the ISP (SLIP/PPP usually).
4. An NT1 device, properly installed and tested at your location.
5. A TA (terminal adapter) device, properly installed, configured, and tested in or attached to your computer.
6. Software to link your computer and your ISDN TA device.
7. A computer.
8. A POTS (analog) telephone or an ISDN (digital) phone.

Part III will explain the procedures involved in obtaining, installing, and getting all of these elements working properly. The rest of this chapter and the rest of Part II will discuss the currently available hardware and software from which you may choose.

WHAT ARE YOUR CHOICES?

Your hardware choices will vary greatly, but they can be divided into three major groups as indicated in Table 5.1.

ISDN Hardware Product Type	Adapter Cards	ISDN Bridges / Routers	ISDN "Modems"
Description	Internal TA or TA/NT1 cards that plug into the Bus and communicate directly (no serial chip use) via TCP/IP, NDIS, ODI or other protocol drivers. They act as network cards.	External network TA or TA/NT1 devices that connect to your computer via an Ethernet cable to your network card, if you have one.	Internal cards or external devices that communicate to your computer via its serial chip, like your analog modem does. They may have TA only, TA/NT1, or TA and analog modem capabilities.
Advantages	Faster throughput than ISDN "modems." May have POTS plug and built-in NT1 at lower price than "modems," bridges or routers. Can use BONDing for 128Kbs throughput.	Faster throughput than ISDN "modems." May have POTS plug and built-in NT1. Can use BONDing for 128Kbs throughput. Easier to setup and configure than adapter cards.	Connect easily to external serial plug or internal as Bus card. Setup easily like a standard modem using standard COM port and existing software. Talk to both analog and ISDN devices.
Drawbacks	Generally difficult to setup and configure, especially with other network cards in same computer. No direct interoperability with analog devices.	Expensive. Require an Ethernet card in your computer. No direct interoperability with analog devices.	Slow performance. One B channel—64Kbs or less with slower UART. No BONDing. May require V.120 to communicate with other ISDN devices.
Examples	Combinet Everywhere Digiboard Datafire ISC SecureLink II ISDN*tek Internet, Commuter, and Enterprise	Ascend Pipeline 50 Gandalf LANLine	IBM WaveRunner Motorola HMTA200

Table 5.1   *ISDN hardware.*

Summary

If you're still with us, you really, really are set on getting yourself deeply into this ISDN thing. Don't worry overmuch about the myriad acronyms, products, and unintelligible jargon. If you can understand what we've presented in this chapter, you can purchase, install, configure, and use your own ISDN equipment on your own computer.

Generally speaking, the adapter cards with built-in NT1 and POTS plug, are the most cost-effective for personal and individual business use—provided that you have the time to hassle with installing and configuring them. ISDN "modems" are the easiest to install and configure, but provide the least amount of functionality at a higher price than the adapter cards. The ISDN bridges and routers are the most expensive, but possibly the most versatile of all three categories: They are reasonably easy to install and configure, providing you already have an Ethernet card in your computer.

If you want to replace your existing POTS phone system with ISDN, you'll want to get a separate NT1 with its own UPS or you'll be without phone service whenever the power goes out. Even then, keeping one POTS line in service is good insurance. In the next chapter, you'll move onward to grapple with the specifics of ISDN hardware and software.

chapter 6

Internal ISDN PC Adapter Cards

ISDN Internal Adapter Cards

ISDN adapter cards are essentially network cards. They plug into your computer and must be installed as network devices, complete with drivers and protocols. This means you must know more than you probably want to know about IRQs, hardware interrupts, I/O addresses, and SRAM buffers. If some or all of this is completely foreign to you, you're not alone. You need to understand these things reasonably well, or else you should get your handy hardware vendor/consultant/local guru to install and configure your adapter card for you.

If you decide to plunge in yourself, be ready to spend hours deciphering your MSD (Microsoft System Diagnostics) or other hardware system configuration program's printouts regarding your system's interrupts and RAM usage while you try to choose among the few remaining available interrupts, I/O base addresses, and scratch RAM buffer spaces for your adapter card. Then, you can get on with configuring your computer with the appropriate network drivers for TCP/IP, presuming you're going to use your ISDN to connect to the Internet, or whatever other protocols you intend to use.

If this looks difficult, you're catching on quickly: It is. There is no such thing as, "plug and play" in the ISDN adapter card field—yet. They're not impossible to install and configure, just difficult at best for the inexperienced networker.

This chapter discusses the most popular ISDN adapter cards available in the first half of 1995. Please keep in mind that many of these cards had been on the market only a few months. When we examined and tested them, only a few users had experience with each card and then only with a few computer systems for a few uses. Since interoperability (which is the buzzword for one ISDN card being able to communicate with another ISDN system) is a primary concern, more experience and testing will be necessary before all of the potential problems surface for each card. All known interoperability problems are discussed with each card below (but it's always the unknown or unexpected ones that get you in the end). These cards are discussed in alphabetical order by vendor name.

For completely up-to-date discussions of ISDN issues, please subscribe to the Internet news group, "comp.dcom.isdn." For the most complete list of ISDN information on the WWW, connect to Dan Kegel's fantastic site at http://alumni.caltech.edu/~dank/isdn/. If you can't find the ISDN information you need from Dan's site, it probably doesn't exist.

The information provided below was current as of June 1995. Prices and availability fluctuate wildly and frequently. Your best bet is to check the WWW sites of each manufacturer, using Dan Kegel's page as a starting place, for price and feature changes when you're ready to buy. For those of you who can't conveniently use the World Wide Web, a complete list of manufacturers along with their addresses and phone numbers is located in Appendix B of this book.

Chase Research—ISDN-PC High-Speed ISDN Adapter Card

The Chase ISDN-PC is a high-performance, multifunction intelligent ISDN adapter card for industry standard PCs running DOS, Windows, or SCO UNIX. Standard software interfaces and AT dialing are presented to the PC, to support off-the-shelf or custom applications.

The ISDN-PC can appear as two independent logical ports, each capable of 115,200 bps simultaneously. These logical ports can be accessed by different host applications or use independently by a single application. Alternatively, the two ports can function as a single aggregated connection capable of up to 230,400 bps. In synchronous interface mode, two logical 64,000 bps ports are offered, or a single aggregated 128,000 bps port (for Internet users, this will probably be the most common configuration).

For the DOS environment, the ISDN-PC offers a standard character-based INT14 interface plus Chase extensions for sending and receiving blocks of data. A Windows 3.x device driver is provided, as well as an SCO UNIX driver. An HDLC frame-oriented interface is used for synchronous mode.

ISDN-PC uses the CCITT V.120 rate adaptation standard to allow communication with devices that run at less than 64,000 bps. V.120 incorporates both error correction and flow control over the ISDN link. The Chase ISDN-PC is fully compatible with the Chase ISDN-TA Fast Terminal Adapter.

The AT dialing interface is presented, allowing easy migration of existing communication applications. Chase extensions have been added to the AT interface to allow control over ISDN services such as calling line ID, multiple subscriber numbering, sub-addressing, and B-channel allocation. These extensions allow more sophisticated applications to be developed, including dial-back and direct dial-in to multiple applications on the same ISDN line.

Synchronous applications can also be easily ported to the ISDN-PC, and the V.25bis dialing protocol is available for SDLC/HDLC applications such as IBM Systems Network Architecture (SNA) or the TCP/IP Point-to-Point Protocol (PPP).

Feature Summary

- V.42bis compression for asynchronous speeds of up to 230,400 bps
- Rate aggregation for maximum throughput
- Standard AT dialing with enhancements for ISDN
- V.120 rate adaption
- Two logical ports
- Synchronous (HDLC) mode with V.25bis dialing
- DOS, Windows, and SCO UNIX drivers
- CLI, MSN, and sub-addressing
- Lifetime warranty

Technical Specifications

- Physical card: 2/3 length ISA bus PC card
- ISDN interface: Basic Rate S-interface 2B+D
- D-channel ISDN protocols: National ISDN-1 (USA), Euro-ISDN, INS NET64 (Japan)

- PC interface: 16-bit ISA bus interface, I/O and memory mapped
- Software interface: DOS INT14 plus Chase extensions for block mode Windows 3.x device driver, SCO UNIX device driver, Synchronous HDLC/SDLC interface
- Logical ports: two independent connections or one aggregated
- Dialing modes: Asynchronous—AT (enhanced for ISDN), synchronous—.25bis
- Rate aggregation: Yes
- Data compression: V.42bis
- Rate adaption: V.120
- Data rates: Asynchronous—300–230400 bps, Synchronous—600–128,000 bps
- Security: Calling line identification
- Other features: Subaddressing, multiple subscriber numbering
- Processor: Motorola 68302
- Physical indicators: LED indicators on card for diagnostics

Comments

The Chase ISDN-PC specifications and features look good, but we have no information on price, availability, software included (if there is any), or user comments. It doesn't appear to include an NT1 or a POTS connection.

Cisco LAN2LAN Personal Office for ISDN

LAN2LAN Personal Office is a remote-node product for ISDN featuring an Open Data-link Interface (ODI) workstation driver (for IPX and TCP/IP) and an ISDN adapter called an ISDN Network Interface Card (INIC). LAN2LAN Personal Office for ISDN is compatible with all Cisco ISDN routers.

The user-friendly AutoInstall utility of the LAN2LAN Personal Office for ISDN eliminates the guesswork from configuration. AutoInstall is so simple that even nontechnical personnel can install and maintain the product.

LAN2LAN Personal Office for ISDN supports calling line identification and the PPP Challenge Handshake Authentication Protocol (CHAP) to ensure a secure internetwork. Using calling line identification, the router that receives the call examines the ISDN number of the caller and verifies whether it is authorized. If the number is not authorized, the call is rejected and no charges are incurred by the calling party.

LAN2LAN Personal Office for ISDN employs DDR technology with the simplicity of PC-client software; thus, this function is handled automatically. During periods of inactivity, the ISDN connection is not active, and no charges are incurred. DDR means you pay for the ISDN connection only when there is data to send. LAN2LAN Personal Office for ISDN and the connected Cisco router achieve DDR by performing intelligent "spoofing" and filtering to ensure that periodic maintenance packets, such as network broadcasts, do not cause the ISDN line to become active.

Features at a Glance

- Hardware and software in a single package
- Integrated ISDN Basic Rate Interface (BRI)
- One ISDN BRI port with RJ-45 connector
- Support of all major ISDN switch protocols
- ODI driver interface currently available.
- Windows NT and Windows 95 interfaces (in future)

Benefits Summary

- Easy to install and configure using the AutoInstall utility
- High-speed solution—56 Kbps and 64 Kbps
- Security to ensure a safe internetwork
- Point-to-Point Protocol (PPP)
- Transparent dial-on-demand routing (DDR)
- Support for the major client platforms

PC Requirements

- 80386 or higher
- Desktop
- At least 1 megabyte of memory
- One available half-length ISA slot
- DOS V5.0 or later

- Windows V3.1 or later
- Windows for Workgroups V3.11 or later

Comments

The Cisco LAN2LAN Personal Office specifications and features look good but we have no information on price, availability, or user comments. It doesn't appear to include an NT1 or a POTS connection. Cisco is a well-known manufacturer of LAN devices, and may well have produced a sound product. It does appear to include an easy-to-use installation and configuration program. Only an ODI driver is currently available, making it unusable for Windows and WfWG users using only NDIS drivers at this time.

Combinet EVERYWARE 1000

Plugging directly into the PC's ISA Bus, the EVERYWARE 1000 looks like an Ethernet adapter card to the PC's software. The EVERYWARE 1000 is compatible with popular network operating systems, including Windows and Novell NetWare, as well as TCP/IP-based applications.

The EVERYWARE 1000 Series is available in the four models listed below, with data compression and integrated NT1 options. With the compression hardware, file transfer rates of up to 512 kbps are achieved on the two aggregated B-channels of the ISDN basic rate line. All models have a DTE-side S/T port for use with an ISDN phone, fax machine, or other ISDN devices.

PC-1030—ISDN BRI PC Card
PC-1040—ISDN BRI PC Card with NT1
PC-1050—ISDN BRI PC Card with Data Compression
PC-1060—ISDN BRI PC Card with NT1 and Data Compression

EVERYWARE 1000 Features

- No separate Ethernet NIC, cables, or hubs required
- Data compression (on some models)
- Built-in NT1 (on some models)
- On-demand dialing

- Supports ISDN phone, fax
- Software upgradable
- Authentication and callback security
- Remote management
- ODI and NDIS drivers included

EVERYWARE 1000 Specifications

- PC bus ISA, single-slot, 2/3 length, 16 bit
- PC desktop compatibility 386, 486, Pentium
- PC software compatibility ODI and NDIS 2.01 drivers
- Compatibility tested with NetWare 3.x, 4.x Windows 3.1, Windows for Workgroups 3.11
- No additional memory required
- Data Compression Stac Hardware, Lempel Ziv Algorithm (models 1050, 1060 only)
- Operates on two B-channels, 128 Kbps (pre-compression) management
- Text-based menu through COM port emulation (9600 baud)
- Combinet Remote Login over Ethernet or ISDN (UDP/IP)
- DIP switches two for COM selection (1, 2, 3, 4)
- ISDN switch types compatibility—AT&T 5ESS, Northern Telecom DMS 100, National ISDN-1 (NI-1),
- NET3/5 (Europe, Singapore, Hong Kong), 1TR6 (Germany), VN3/4 (France)
- ISDN Line Interfaces ISDN BRI S/T RJ-45 (Models 1030, 1050); or ISDN BRI U RJ-45 (Models 1040, 1060)
- Compatible with data, voice, or voice/data B-channels
- Multipoint configuration
- ISDN Terminal Device Interface
- S/T port for ISDN phone/fax support

EVERYWARE Software

Combinet EVERYWARE software is included with each Combinet on-demand networking access unit and can be upgraded simply by downloading a diskette file. EVERYWARE access units give users remote access to all applications and data resident on an enterprise network. Combinet Software benefits include the following:

- Software upgradable for investment protection
- Simple installation and maintenance
- Interoperable with all Combinet products
- Efficiently manages telephone connections to minimize usage charges
- Compatible with standards-based ISDN BRI, Switched-56 telephone service, and Ethernet LANs
- Computer, network operating system, and application transparency
- Extensive security options

Comments

Combinet's ISDN bridges and routers have been solid industry workhorses for several years. It can be hoped that the EVERYWARE 1000 carries on this tradition. None of the models seems to include a POTS plug, and the compression is only useful if it is compatible with the compression technique used by the device on the other end of your ISDN connection, and, the device has compression turned on.

DigiBoard—DataFire Personal High-Speed Remote LAN

DigiBoard's DataFire adapter card integrates an ISDN terminal adapter/network interface card with a network terminator (NT1) to provide simplicity and affordability in ISDN remote LAN connections for individual computer users. The DataFire installs in an expansion slot of ISA-bus PCs and connects directly to an ISDN Basic Rate Interface (BRI) phone line.

One 128 Kbps or two 64 kbps communications channels are supported. The DataFire integrates with Microsoft Windows NT, Windows for Workgroups, and Novell NetWare. DigiBoard's Link Optimizing Software reduces line charges by up to 60 percent. The use of PPP (Point-to-Point Protocol) support ensures interoperability with other types of ISDN devices, including most Internet service providers and other online resources.

Features/Specifications

- Supports two B channels and one D channel; B channel voice or circuit-switched data connections
- Supports "nailed up" permanent circuit-switched B channel
- Compatible with 64 Kbps, 56 Kbps and ISDN voice circuit-switched calls
- Supports Q.931 and Q.921 (LAPD), HDLC, and proprietary protocols on voice and circuit-switched data channels
- Compatible with all CCITT/ANSI BRI ISDN central office switches, including AT&T 5E, Northern Telecom DMS 100, Siemens, and British Telecom
- Connects between ISDN switches using digital dial-up voice lines and digital trunking
- Data compression for greater throughput
- Point-to-Point Protocol (PPP) support for interoperability
- Transparent connection management—no dialing commands necessary
- English language configuration and control
- Programmable operation
- Automatic fallback from circuit-switched 64 Kbps calls to receive 56 Kbps calls or send at 56 Kbps over ISDN voice calls
- ISDN U interface accepts RJ-45 or RJ-11 connectors
- Supports client LAN software for Novell, Microsoft RAS, and Internet PPP server types
- Supports DOS ODI 3.11 and DOS NDIS 2.0
- Supports Windows NT 3.1, 3.5, Windows for Workgroups 3.11

Comments

DigiBoard's DataFire is one of the first combined TA/NT1 adapter cards available for PCs. Its low price (list $595) has made it desirable for individuals wanting to get onto the Internet via ISDN. Unfortunately, the DataFire has no POTS connection (but the ISC SecureLink card does). Also, at least one owner of a DataFire has been unable to make the DigiBoard-supplied drivers work with his NT 3.5 system, even after more than a month of trying with help from DigiBoard's customer support personnel. DigiBoard makes good products, but it appears that the DataFire or at least some of its driver software isn't quite ready for prime time. Nor are the card's features up to the level of at least one other card on the market at about the same price (the ISC SecureLink).

Euronis Gazel-ISDN Board

The Gazel card works in PC-AT based microcomputers that have an ISA 8- or 16-bit bus. It interfaces with MS-DOS, Windows, Novell NetWare, and UNIX operating systems. The Gazel board (also known as the Gazel-2B board) offers the ability to assign the two B channels to data transfer. This enables simultaneous communication with two different sites at 64 kbps or with a single site at 128 kbps. Rate adaptation is also supported for 56 kbps connections.

Working as a background task on your PC, the "Gazel Phone" is a telephony software application that manages incoming and outgoing voice calls on one of the B channels. Among its features you'll find a built-in directory.

The Gazel board works on ISDN lines originating from AT&T 5ESS, Northern Telecom DMS-100, or Siemens EWSD central office switches (both custom and National ISDN 1 lines).

Technical Specifications

- Half-size card for ISA bus PCs
- ISDN Basic Rate, 2B+D
- RJ-45 socket for connection to Network Terminator (NT-1)
- RJ-11 socket for connection of standard phone
- NetBIOS Driver
- Communication protocols: HDLC, X.25 64kb/sec, X.25 128 Kb/sec (2B channels), Multi-B channel X.25 for nx64 connections.

Software Included

- STARTX, software controlling the Gazel card via the AT bus, provides applications with programming interfaces based on NetBios
- XFER, a simple file transfer utility

- SPYX and SPYRNIS, B channel and D channel symbolic trace tools for application developers
- PHIDOS and PHIWIN, voice telephony software for MS-DOS and Windows.

Comments

The Gazel-2B board was developed for the European market where the NT1 is part of the ISDN service from the local phone company. The card does include a POTS plug, however. Although it may be a good card, getting support from Europe might be less than attractive for most U.S. customers. You had better check this one out carefully before jumping in.

Intel—RemoteExpress ISDN LAN Adapter

According to Intel, the RemoteExpress adapter is a ready-to-use solution. Board configuration is automatic, and there are no switches or jumpers to set and no IRQs. The RemoteExpress adapter comes with built-in drivers for most popular networks. In addition, three levels of security protect the network every time a remote computer tries to log on. First, a unique MAC-level address identifies the remote user's personal RemoteExpress adapter, making sure it is registered and authorized. Second, the remote user PC name protects against entry by unauthorized computers. And third, the network password must be supplied correctly.

Requirements
- ISDN BRI S/T phone line (available through local phone company)
- Intel 386 SX-based computer or higher, 4MB RAM, DOS 5.0 or higher
- NT1 with power supply
- Product: RemoteExpress ISDN LAN Adapter
- Order Number: PCIS9500

Prices

Product	Order Number	Price
RemoteExpress™ ISDN LAN Adapter	PCIS9500	$499
BellSouth region (5)	PCIS9500	$399
GTE region	PCIS9500	$199
NYNEX region	PCIS9500	$199
Pacific Bell region	PCIS9500	$199
Other regions	PCIS9500	$499
RemoteExpress ISDN Bridge Pack	PCIS9501	$2,199
BellSouth region (5)	PCIS9501	$2,199
GTE region	PCIS9501	$2,199
NYNEX region	PCVD9501	$2,199
Pacific Bell region	PCIS9501	$2,199
Other regions	PCIS9501	$2,199

Comments

Marketing relationships between Intel and the telephone companies listed above may qualify you for lower prices when purchasing RemoteExpress products in conjunction with new ISDN service. Contact an Intel Advanced Network Reseller for details. Or, for a list, request FaxBack document 8212. To find out if there's ISDN service in your area, call Intel at 1-800-538-3373, ext. 208. Currently not available outside of North America.

Little information is available from Intel about this product; however, according to ISDN Systems, Inc., the RemoteExpress card is Intel's version of the ISC SecureLink ISDN adapter card, discussed next.

ISDN Systems, Inc. ISC SecureLink ISDN Adapter

The SecureLink I ISDN Adapter was designed to work with the same familiar software used in the office LAN environment. The SecureLink I includes NDIS, ODI (for IPX), Packet Driver, and TAPI drivers so that it can be used with the most popular, off-the-shelf software packages.

The SecureLink I ISDN Adapter is equipped with an RJ-11 (POTS) jack so you can use your analog telephone to communicate over the digital network. Dynamic bandwidth allocation, compression, PPP, multilink PPP, integrated NT1, RJ-11 analog phone jack, Internet access, security via SecurID card, and TAPI provide a complete solution for a wide variety of communications applications.

When used in conjunction with the SecureLink I ISDN Server, the SecureLink II Adapter offers unparalleled security. Multiple levels of security protect your LAN from unauthorized access. For an extra measure of security the SecureLink II works in conjunction with the SecurID card from Security Dynamics.

Shipped with Card, Quick Start Up Guide, Reference Guide, RJ-45 cable, SecureLink driver installation disks, and Microsoft Windows Telephony (TAPI) installation disk. Price: $199 to $595.

Requirements
- ISDN BRI phone line (available through local phone company)
- PC 386 CPU-based or higher, 4 MB RAM, DOS 5.0 or higher

Specifications
- PPP and Multilink PPP
- TCP/IP or IPX routed
- NetBEUI, 802.3 or 802.5 bridged
- All major LANs supported
- ISA or EISA
- DOS
- Windows 3.1 & 3.11
- Windows 95 and Windows NT 3.5
- TAPI on-screen dialing
- Simultaneous voice/data
- 2 S/T interface plugs
- U interface
- RJ-11 for POTS

- Built-in NT1
- Inter-switch at 56K
- Single B at 56K
- Combined 2 b (112/128)
- V.120
- NDIS, ODI (NetWare), Packet Driver, TAPI, WinISDN
- Bandwidth on demand
- Compression
- SecurID from Security Dynamics
- Softset card (no jumpers)

Comments

The SecureLink ISDN adapter card plugs right into your computer and contains all of the functions you could want from an ISDN adapter card priced under $600 retail. It comes with professionally written user's manuals and well-thought-out installation, driver, and utility software. The ISC support personnel are very responsive to customers' needs and will even call you back to see if everything is working well. They were more experienced with Novell and ODI drivers than with Windows and NDIS systems, but they learned quickly.

The original SecureLink cards sometimes had problems with shared SRAM address space in the computers in which they were installed, so the new SecureLink I cards contain onboard RAM space, which is supposed to remedy this problem. Personal experience has shown that the SecureLink card works well in a WfWG environment, once the SRAM conflicts are diagnosed and avoided. The installation and use of this card is the focus of Chapter 11 of this book. This is one of the best buys in the ISDN adapter card market today.

ISDN*tek CyberSpace Freedom Series Internet Cards

All the CyberSpace Freedom Series (Internet, Commuter, and Enterprise) adapter cards provide a WinISDN driver that can be addressed by application software through the industry standard WinISDN interface. Cards are available for Internet access, for telecommuters, and for demanding enterprise environments. An Integral NT1 version is also available for each type of card.

CyberSpace Internet Card

The CyberSpace Internet Card is designed to provide low-cost ISDN access to the global Internet, with simplicity of installation and reliability as key features. The WinISDN Application Program Interface (API) provides Windows- and WinOS/2-based PCs with an easy standard interface to ISDN-ready TCP/IP software. The Internet is then accessed through routers maintained by service providers. Because the typical Internet access does not support voice calls, the Internet card does not possess voice call capabilities. The CyberSpace Internet card at $395 supports a 56 Kbps or 64 Kbps data call on either B-channel and requires an external Network Termination (NT1). It can share the ISDN line with other ISDN hardware.

The CyberSpace Internet+Plus card features an on-board NT1 at $495 and can be connected directly to the phone company's wall jack. This configuration eliminates the external NT1 and dedicates the ISDN line to this card. The card also supports either one-channel or two-channel data calls. When used with appropriate multilink PPP software, the user can obtain up to 128 Kbps data rates.

Features

- PC (ISA-bus) hardware
- Windows drivers
- 56K/64K data call over one B-channel
- 128K data call over two B-channels using MPPP software
- Compatible with Internet Software and routers
- Versions for "S" or "U" ISDN connections
- Japanese INS64 version available

CyberSpace Commuter Card

The CyberSpace Commuter card is designed to provide a fully digital telecommuting solution. The Commuter card supports both B-channels of the ISDN connection, which can be used for voice and data or for up to 128K bps of data. It retains all the features of the Internet card, in that it can support a 56K to 64K data call on either B-channel, while data on both B-channels can be aggregated to 128K bps using appropriate Multilink-PPP (MP) software. The card continues to support Internet access software and to communicate with Internet routers.

The significant new feature of the Commuter card is voice call capability. Not to be confused with analog voice service, the commuter card uses a standard headset or handset, thereby eliminating the need to purchase an expensive digital telephone. The telephony functions normally found on the telephone keypad are handled by the supplied dialing application running on the computer. The telephone line can be configured so that when the computer is not powered, incoming calls can automatically be routed to a telco-maintained voice mail system or back to the corporate office.

The Commuter Card requires either a standard handset or an amplifier equipped headset to support voice. It is a 100% digital ISDN product and does not have an analog phone port and does not support fax, modem, or analog phone sets. It does, however, include the ability to send touchtones for accessing automatic call distribution (ACD) and credit card data entry equipment.

The CyberSpace Commuter card at $495 requires an external NT1 and can be used with other ISDN equipment. The CyberSpace Commuter+Plus card at $595 includes an on-board NT1, which can be connected directly to the phone company's wall jack. Using this configuration with a headset eliminates all other external equipment such as an NT1 and desktop phone set and dedicates the ISDN line to this card.

Features

- PC (ISA-bus) hardware
- Windows drivers
- 56K/64K data call over one B-channel
- 128K data call over two B-channels
- Voice call over one B-channel.
- Simultaneous voice and data calls over the two B-channels
- Sends touchtone DTMF while call is active
- Streaming data mode for rerouting voice
- Voice capture mode for answering machine applications
- Versions for "S" or "U" ISDN connections

- Four-wire headset or handset compatible voice jack
- Compatible with Internet software and routers

CyberSpace Enterprise Card

The Enterprise card is a superset of the Commuter Card and offers several logical channels of X.25 connectivity over the D-channel of the ISDN Basic Rate Interface. Voice and data connections are similar to those of the Commuter Card. In addition, the Enterprise card supports DTMF "touchtone" detection so that incoming calls can be automatically routed or so that recorded messages can be retrieved remotely from disk.

The X.25 or packet connections can be used for sending small messages or control signals from the computer to remote equipment. Simple messages to the host system or another peer, such as a code to call back with file information on a customer, can be sent while both B-channels are active with other duties. This allows an incredible variety of network connections for enterprises, particularly those "home-based" enterprises of the 21st century.

Features

- PC (ISA-bus) hardware
- Windows drivers
- 56K/64K data call over one B-channel
- 128K data call over two B-channels
- Voice call over one B-channel
- Simultaneous voice and data calls over the two B-channels
- Several logical channels of X.25 over D-channel
- Generates and detects touchtone DTMF while call is active
- Supports call hold, conference, and transfer
- Streaming data mode for rerouting voice
- Streaming video mode
- Video capture for surveillance
- Voice capture mode for answering machine applications
- Supports ISDN "U" connections

- Four-wire headset or handset compatible voice jack
- Compatible with Internet software and routers

The Driver Software

The WinISDN.DLL driver supports synchronous PPP (Point-to-Point Protocol) and HDLC from the Windows-based TCP/IP stack mentioned earlier. It works with service providers' ISDN routers when used with any WinISDN-based PPP software. It supports FTP, Ping, E-mail, Gopher, Mosaic, etc., when used with appropriate TCP/IP software. Driver support for peer-to-peer connectivity, messaging, file transfers, streaming data, and voice connections is also included. The voice product includes a dialer/phonebook mini-application. A Software Developer's Kit for easy interface to Visual Basic or C programs also is available.

Comments

ISDN*tek's six offerings are professional-looking, including the user manual and packaging. They come complete with installation and testing software. You have to set dip switches on the card, but the installation software runs on your computer and helps you determine what settings will be required before you plug the card into your computer.

The use of a regular telephone handset or headset with the Commuter and Enterprise cards to achieve digital telephony is an added plus. The lack of NDIS and ODI drivers for the cards and the lack of Windows NT drivers will make many potential users skip these products in favor of others until the drivers are available. If you have the correct hardware to use the ISDN*tek cards, they are a good buy and should be considered as reasonable competition for the ISC SecureLink II card.

Lion DataPump

The DataPump is an IBM-PC compatible half-card ISDN TA and digital or analog fax/modem. It allows calls to be made from an ISDN line to conventional modems and faxes. It is also an ISDN TA adapter for 64 Kbps, with reverse multiplexing 128 Kbps data rates. It also can be used without an ISDN line: By means of an adapter cable assembly with a line interface module, it can connect to a normal telephone line and will then work as a conventional fax modem at 9600 bps or a V.22 bis modem at 2400 bps. With the built-in V.42bis, however, V.42 data rates of 9600 bps can be achieved. (also includes MNP 2–5).

With an ISDN-to-ISDN connection, or ISDN-to-analog, or analog-to-analog, it uses the AT Hayes command sets and will work with normal communication software accessed via the COM ports. Therefore, if you have conventional software for analog modems, the DataPump will work with it.

The card is full Euro-ISDN card compatible (ETSI ETS 300 102-1). It also includes the VN2 French ISDN standard protocol and the old German National ISDN standard 1 TR 6. It also conforms to the UK BTNR191 standard and CCITT Q931 standards.

Specifications

- ISA direct bus interface—it uses the S0 interface (S/T), and thus up to eight such devices can share this interface.
- The ISDN connector is the I.420 also called RJ45, ISO8877, FCC68/8.
- The data protocols handled over ISDN to an analog line are CCITT V.29, V.27 ter, V.22 bis, V.22, V.21, and Bell 212A, 103. It supports synchronous and asynchronous mode SDLC, HDLC Bisync covering data rates from 9600 bps to 300 bps.
- The compression & error protection protocols are: CCITT V.42, V.42 bis, and Miracom MNP 2-5.
- Its fax protocol is Group 3 Class 1 & 2 EIA/TIA TR29.2.
- Its ISDN protocols are V.14, V.110 (56K bps synchronous and asynchronous), V.120, X.75.
- Point to Point Protocol (PPP).
- Bonding 1.1.
- Common Application Protocol Interface (CAPI) 2.0.
- Price: The DataPump costs DM599 ($400 US).

Comments

The DataPump seems almost too good to be true. It's available directly from Lion in the UK since they don't seem to have any U.S. distributors yet. It is obviously designed for the European market with no NT1. If you need all of the analog support it provides and want to use a separate NT1, it may be of interest to you. Also, if you don't mind buying it from a company across the pond with the support problems that entails, it may turn out to be a good solution.

Surf Communications, Inc.—ExpressPak

Surf Communications says the ExpressPak is a complete ISDN Internet Kit for Windows. It contains a high-speed ISDN card with NT-1, TCP/IP stack, e-mail (Eudora), FTP, and Surf's offer to order and test your ISDN phone line for you. ExpressPak is available direct from Surf Communications for $399.

Any application that works with your existing modem dial-up Internet account will work with ExpressPak. ExpressPak is compatible with other ISDN cards and routers. ExpressPak currently supports two B channel BONDing and will support MultiLink PPP via a software upgrade by mid-1995.

Installation is very straightforward: The ExpressPak ISDN card plugs into a open ISA slot in your PC. You then connect the ISDN line with the included RJ-45 cable. The software is installed using the ExpressPak Installer. All operating software, drivers, and applications are loaded and preconfigured.

Surf Communications, Inc. provides unlimited tech support for your ISDN connections and your Internet account. You can contact them via their voice support line or via e-mail. Their support starts with ordering your ISDN line, making sure it's installed (and provisioned) correctly, and testing the line. To insure a trouble-free installation, they strongly suggest letting them order and test your ISDN line for you.

The ExpressPak card has a standard POTS phone jack that supports an analog phone. It also dynamically allocates the B channels—that is, if you are transferring data on both B channels and pick up the phone, the card will automatically drop down to one B channel for data and one for phone. You can also receive a normal incoming phone call on your second B channel, if you are not using it to transfer data. You can plug in a fax or modem into the same jack; however, they do not support this use. Keep in mind that your computer must be turned on to activate the NT1 and use the POTS phone jack.

Comments

Surf doesn't provide specifications on the ExpressPak adapter card on its WWW server, probably because they are concentrating on turnkey Internet service to their primary customer base in California. They are interested in expanding their service and may be interested in selling the ExpressPak to people who only want the adapter card. We included their information to show what some ISPs are doing for their intended customers. Their idea of ordering your ISDN service for you and testing the line themselves should be comforting to many anxious ISDN clients.

SUMMARY

As you can see from the impressive group of ISDN adapter cards, you need not be limited by a lack of good hardware choices. Of these cards, the ISC SecureLink I and Intel RemoteExpress (which is the same card as the SecureLink I) are the front runners in hardware functionality, software, and relative ease of installation. The ISDN*tek group of (Internet, Commuter, and Enterprise) cards runs a close second, trailing only because of their lack of drivers for NDIS, ODI, and Windows NT.

To reiterate, none of these cards is plug-and-play. All are network cards that require installation of drivers and configuration of your computer's I/O addresses, IRQs, and usually an SRAM buffer. That's the nature of network cards. They are more difficult to install and configure, but they offer the best performance for the price after you get them up and running.

No doubt every manufacturer is frantically attempting to make its card plug-and-play. But it will probably be quite a while, if ever, before one of them actually writes software that can properly dissect the intricacies of a given PC's hardware and software and configure the adapter card accordingly. It's a bit of an AI problem that's much more difficult than it seems to the inexperienced user.

Our best advice regarding ISDN adapter cards is to be sure you really want their benefits, despite their current costs, and their drains on your time and sanity. Part three of this book provides detailed instructions on setting up a home office using an ISC SecureLink card, as well as case histories of other installations. If you read Chapter 11 and still want to install an ISDN adapter card yourself, you're either as crazy as we are, or you know more about ISDN than we did. Either way, you'll probably succeed. Good luck!

chapter 7

ISDN Modems and NT1s

ISDN "Modems"

If your guiding principle is KISS (Keep It Simple, Stupid!), then don't ignore the simplicity of so-called ISDN modems. While they may not be quite as fast nor as loaded with features as ISDN adapter cards, or as ISDN bridges and routers, they certainly can produce the desired results. And they're a lot easier to use.

Some of these ISDN modems are as easy to install as plugging them into your wall power outlet, jacking an RJ45 cable (supplied with the unit) into your ISDN wall jack, and connecting the "modem" to your serial port with a standard DB25 cable (not always included). Some can be configured by plugging your POTS phone into the RJ11 jack on the "modem" and setting it via the telephone keypad. Others offer DOS or Windows setup programs.

All of the ISDN modems avoid the major installation and configuration problems found with the ISDN network adapter cards: That is, they let you avoid the problems common to setting shared memory buffers, I/O addresses, IRQs, and network drivers. To your computer, the ISDN "modem" is just that: another modem accessible through the RS232 serial port and its associated UART (or even through its own UART, as is the case of the IBM WaveRunner).

The best of these "modems" use ISDN standard protocols when communicating with other ISDN devices and use standard Hayes AT commands for communicating with your computer and other analog devices. Using one of these devices is really like using a much faster modem. In fact, some of them can even sense the type of incoming call and use the appropriate analog or digital protocols automatically.

All of these ISDN "modems" (except for the IBM WaveRunner) are subject to the constraints of your computer's serial UART chip. If you're using a newer computer with a 16550A UART, you may be able to attain transmission rates of 57.6 Kbps, otherwise you'll be limited to about 19.2 Kbps. If you have an older computer, the IBM WaveRunner ISDN modem contains its own high-speed serial chip to avoid that problem on your behalf. None of the ISDN modems on the market currently support two B channel BONDing, so you can't even get the theoretical 115 Kbps speed of the ISDN line using the serial device.

On the positive side, all of the ISDN modems—except for the IBM WaveRunner—are external devices, so you don't have to uncase your PC hook them up. Some include built-in NT1s, while others don't. In either case, they all are powered by their own power cables that draw power from your AC wall outlet. This means that if the ISDN modem has a POTS jack (two in the ADAK 221) for your analog phone, you can use it with your computer turned off. Some even have an internal battery backup in case your electricity goes off. Of course, the ISDN modems with internal NT1s and battery backup are more expensive, but the extra features may be worth the additional costs, depending on how you plan to use them.

Several high-quality ISDN modems are discussed in this chapter. All work well, but no two are completely equivalent because they all have different feature sets. Since the industry is still in its infancy, standardization is nonexistent; since each of these new products is trying to carve out a new market niche, every manufacturer is constantly trying to "out-feature" the competition.

You would be wise to check the latest trade magazines and Internet news groups (such as *comp.dcom.isdn* or *comp.dcom.modems*) before purchasing an ISDN modem. You should also check with your Internet service provider and local telephone service provider to make sure your choice works with their systems. If it doesn't, it won't matter how much you like the equipment; if it doesn't work, you'll have to find something else!

Network Termination 1000 Devices (NT1)

If the ISDN modem doesn't contain an internal NT1 (Network Termination 1000 device), you will need to purchase one separately. NT1s range from small, simple-looking devices with a couple of RJ45 jacks (one for a cable to the ISDN wall jack and the other to your T/A (adapter card or ISDN modem) and power cord jack (if it doesn't have its own power supply), to the full-featured IBM 7845 NT1 Extended with its programmable interface and multiple S/T and analog output jacks. Some NT1s are designed to be used with external UPSs. These are very useful when you have several analog and digital devices that need to run continuously, even when your computer is off.

Choosing a good NT1 is much simpler than choosing an ISDN T/A. Check with the manufacturer of your chosen T/A (adapter card or ISDN modem) to insure compatibility with the NT1, but virtually all are interoperable. Whether the NT1 has a battery backup or a UPS for power failures may be very important to you as well. If you are simply looking for an external NT1 to link your ISDN modem to your ISDN line with no analog (POTS) phone jack in the NT1, purchase the simplest and cheapest. If you want more features, look at the IBM 7845 or others that will undoubtedly be on the market soon, or consider getting an ISDN bridge or router with a built-in NT1 that offers the features you need.

All of the NT1s discussed in this chapter should provide good service if you follow the checklists and purchase the model that suits your needs. Their manufacturers are reliable and stand behind their products. Always double-check their warranties and keep your receipts, however: Most vendors require proof of purchase before they're willing to admit they owe you anything at all!

Meet the ISDN Modems

3Com Impact ISDN External Digital Modem (QuickAccess Remote External Digital Modem)

The 3Com Impact™ ISDN External Digital Modem utilizes ISDN services available from the telephone company to transmit at 56 or 64 Kbps—about

25 times faster than 2.4 Kbps modems, and at least twice as fast as the latest generation of V.34 modems. The 3Com Impact modem is compatible with a wide array of Internet browsers, communications software packages for modems, and Internet Service Provider (ISP) equipment.

Through the AT command set and the IETF's Point-to-Point Protocol (PPP), the 3Com Impact modem supports all major Internet browsers. The 3Com Impact implements V.30 and Async-Sync(TM) PPP conversion, and it has been tested for interoperability with ISDN equipment in use at leading ISPs. Its analog voice port supports existing telephones, fax machines, and answering machines, letting them operate simultaneously with PC data transmission and allowing a home office user to replace several analog lines with a single ISDN line. To avoid the need for extra equipment, ISDN network termination (NT-1 for the ISDN U interface) is integrated into the 3Com Impact. Front panel LEDs and a diagnostic self-test give the user valuable information on ISDN line status and connections.

Specifications

- Length: 10 in/25.4 cm
- Width: 7 in/17.8 cm
- Height: 1.5 in/3.8 cm
- ISDN Basic Rate Interface (BRI): Two 56 or 64 Kbps bearer (B) channels
- and one 16 Kbps control (D) channel
- Built-in NT-1 termination with 2B1Q interface
- North American National ISDN-1 (NI-1)
- AT&T and Northern Telecom custom ISDN signaling
- Circuit switched data
- Circuit switched voice
- Support for switched 56 circuits
- Compatibility with switched 56 CSUs and ISDN terminal adapters (TAs)
- TCP/IP PPP
- Async PPP to sync PPP conversion
- ANSI v.120
- µ-law pulse-code modulation (PCM)

- Voice and G3 fax
- Data Port—female 9-pin D connector
- RS-232 at 115.2 Kbps asynchronous
- AT command set dialing
- Autobaud matching of transmission speeds up to 115.2 Kbps
- Hardware flow control
- Optional voice port—RJ11 modular connector
- Touch-Tone (AT&T 2500) compatibility
- Battery feed: 20 mA
- Ringing voltage: 86 VAC
- Two extensions within 100 feet (30.5 meters)
- LEDs for operation, call status, and local diagnostics
- Remote diagnostics
- Loopback tests
- Flash memory for field software updates
- Mean time between failures 50,000 hours
- Power—Input: 117 VAC
- Output: 12 VAC
- Power dissipation: 8 W
- QuickAccess Express Digital Modem (includes user manual and quick installation guide, wall power supply, serial cable, and software diskettes)
- For PCs, the model number is 3C870
- For Macintoshes, the model number is 3C875
- VMS DOS configuration software (includes TurboCom Windows Software and Internet Chameleon 30-day trial copy) 3P871
- Macintosh configuration software 3C875
- List price $495.00 base unit
- Voice option: $200.00

Comments

The 3Com Impact ISDN External Digital Modem and QuickAccess Remote External Digital Modem are the same since 3Com purchased AccessWorks

Communications Inc., makers of the QuickAccess Remote. The device contains an internal NT1 and POTS jack along with an impressive set of features and software. At about $700, it may be a good buy in this category.

ADAK 220 and 221

ADAK Communications Corporation's ADAK 220 and 221 combine advanced call management capabilities with high-speed digital communications. The ADAK 221 has an S/T digital-out port that supports ISDN S/T data based and video adapters used for video teleconferencing.

Both units allow access to the global X.25 network, offering cost-effective data communications. The ADAK 220 and 221 support ISDN National-1 (NI-1), National-2 (NI-2), and custom central office switch software. Both ADAK units contain two POTS jacks, two RS232 plugs, and a U interface plug. The 221 also sports an S/T output jack for other ISDN devices.

Specifications
- ISDN Service Options [Basic Rate Interface (BRI)]
- Integrated NT-1 functionality
- Integrated terminal adapter functionality
- Automatic POTS to ISDN service cutover
- Backup power (8 hr) (optional)
- Performs incoming call acceptance routing or rejection via Caller ID
- A standard analog Caller ID box or display telephone can be connected to the analog lines
- Call hold, forward, conference, transfer, screening, and multiple call appearances
- Ability to activate central office switch capabilities (e.g., voice mail, call forwarding, call conferencing); "N" line key phone emulation
- Audio response and prompts for user interaction
- X.25 PAD access over the D channel (X.3, X.28, X.29, and T3POS)
- Hayes compatible modem PAD for access to X.25 network
- Support for security/alarm applications (optional)
- Support for point-of-sale applications
- Compatible with Siemens, Northern Telecom, and AT&T switches

- National-1 (NI-1) and AT&T custom
- Two standard RJ14C analog connectors
- Supports up to five ringer equivalencies on each line for analog telephones
- Two RS-232 serial ports supporting synchronous data speeds up to 128 Kbps and asynchronous speeds up to 57.6 Kbps (DB25 ports)
- Reverse charging capability on data calls
- ISDN BRI "S/T" Port
- An RJ45 "S/T" port supports two 64 Kbps B channels and one 16 Kbps D channel in multipoint mode (ADAK 221 only)
- Battery-backed "Power One" supplied continuously at 1 watt, and "Power Two" supplied with wall power up to 8 watts
- One RJ45 ISDN BRI for the "U" reference point
- Configuration is implemented through a standard analog telephone interface and voice prompts with keypad input (DTMF) or serial device (PC or terminal)
- Local or remote configuration
- LED status indicators—Power, two error, and network activity/status indicators
- Software can be upgraded via RS-232 or remote D channel software download
- Support for remote configuration
- Extensive remote and local internal diagnostics, including complete testing of the ISDN line
- Dimensions: 9.5" x 9.3" x 3.1"
- Weight: 3 lb. (w/optional battery: 4 lb.)
- Integrated with a surge suppressor.
- Includes a UPS with an optional internal battery to supply 8 hours of backup time

Description	List Price
ADAK Model 220	$949.00
ADAK Model 221	$999.00

Backup Battery	$29.99
Standard cable	$17.99
Description	**Extended Warranty and Software Upgrades/Yr.**
ADAK Model 220	$69.00
ADAK Model 221	$79.00
Miscellaneous	
Optional Protocols	$75.00
Special Features (CSD)	$50.00
Software Upgrade Annual Fee	$40.00

Comments

The impressive list of features and known dependability of the ADAK 220 and 221 make them slightly expensive but solid contenders in the ISDN modem with built-in NT1 category. They are especially impressive if you need the extra POTS jack and the optional 8 hour battery for the built-in UPS. The 221 will run all necessary communications devices, analog and digital, for a serious single-person office.

Adtran ISU Express

The ISU Express (Table 7.1 on page 137) looks and feels like an analog modem, but uses a high-speed ISDN line. This unit is a scaled-down version of the ISU 128 specifically designed as a low-cost solution for applications such as high-speed modem replacements, telecommuting, and remote office interconnections. For remote office applications, the ISU Express is available with an optional telephone interface for an analog telephone and/or a fax machine connection.

The ISU Express supports an optional integrated V.32 bis/V.42 bis modem for analog modem interoperability. The ISU Express transmits data over an RS-232 interface and performs at synchronous data transfer rates to 128 Kbps and asynchronous rates to 115.2 Kbps. At rates over 64 Kbps, the BONDing delay equalization protocol synchronizes data over the two 64 Kbps B channels. ISDN network termination is designed into the

Model	Number
ISU Express, data only	1200081L1
ISU Express, with V.32 bis/V.42 bis modem option	1200081L2
ISU Express, with POTS telephone interface	1200081L3
ISU Express, with POTS interface and modem	1200081L4

Table 7.1 *ISU Express model numbers.*

unit, eliminating the need for and expense of an extra NT1. For interoperability, the ISU Express supports many rate adaptation protocols including Clear Channel, CCITT V.120, BONDing, T-Link, ADTRAN DSU 57.6, Fallback, and Async-to-Sync PPP. Configuration of the unit is achieved using a VT 100 menu system and AT commands. Dialing is initiated in-band over the RS-232 interface using AT commands, V.25 bis, or DTR assertion. The front panel contains LED indicators for power, network readiness, DTE interface leads, ring indication/off hook status, and local and remote loopback testing.

Comments

The Adtran ISU Express looks good based on its features and its spec sheet, but there isn't enough user data or testing information available for any meaningful comments yet. If you find it intriguing, please check the newsgroups and ask your ISP before you part with any money for one.

IBM WaveRunner Digital Modem with Enhancements

The IBM WaveRunner Digital Modem is a PC adapter card that looks like a standard analog modem to your computer, but provides BRI connectivity for communications over an ISDN network. The unique feature of this adapter is its ability to communicate through ISDN with devices such as PCs and fax machines attached to an existing analog telephone line via modems. This function allows end users to have a mix of devices distributed over ISDN and analog lines, while maintaining an ability to communicate from devices attached to an ISDN line with those on an analog line. The adapter also supports many existing, widely used modem application programs at speeds up to 64 Kbps over ISDN and up to 14.4 Kbps (without compression) on analog modems.

The WaveRunner supports ISDN Basic Rate Interface 2B+D under the Microsoft Windows 3.1 environment on networks conforming to National ISDN-1, Custom (AT&T 5ESS, 5E7, 5E8, and 5E9 switches), Euro-ISDN 1TR6, and VN4 standards, and 1B+D under OS/2 environment on networks conforming to National ISDN-1 and Custom (AT&T 5ESS, 5E7, 5E8, and 5E9 switches) standards. Full support for the OS/2 environment for Euro-ISDN, 1TR6, and VN4 standards is also available, but not so widely deployed.

Features

- Integrated Services Digital Network (ISDN) Basic Rate Interface 2B + D channels, in data mode, at speeds up to 64 Kbps per B channel
- Network Driver Interface Specifications (NDIS)
- Packet level BONDing for IP packets to provide a total data rate of 128 Kbps without compression, when using both B channels
- Common Application Programming Interface (CAPI)
- WinISDN API for Microsoft Windows
- Data compression of up to 4:1, using V.42bis when communicating with modems over ISDN
- Data compression of up to 4:1 when using V.42bis over V.120, at 56 Kbps or 64 Kbps, when communicating with other ISDN devices
- X.25 on B and D channels when used with IBM Communication Manager/2 in an OS/2® environment
- Euro-ISDN, 1TR6, and VN4
- Supported under Microsoft Windows 3.1 on networks conforming to National ISDN-1, Custom (AT&T 5ESS with 5E7, 5E8, and 5E9 switches), Euro-ISDN, 1TR6, VN4 standards, and OS/2 environment on networks conforming to National ISDN-1, Custom (AT&T 5ESS with 5E7, 5E8, and 5E9 switches) standards:
- COM Port Type I and Type II serial port emulation
- Permits use of existing PC modem communication software over ISDN
- Provides a 2:1 compression for a significant increase in throughput by using MNP 4/5 data compression on modems

- Hayes AT® command set
- When used with the IBM LAN Distance set of software products, allows a stand-alone workstation configured with the COM Port MAC driver to access remote LAN for all available LAN services
- All commonly used PC modem communications programs are supported
- Fax emulation (for Windows 3.1) enables PCs to exchange standard Group 3 facsimile data with fax machines connected to an existing analog telephone network when the WaveRunner running under Windows 3.1 is used with the fax application program shipped with the adapter
- CAPI support permits the use of ISDN application programs written for CAPI
- WinISDN API for Microsoft Windows
- Provides access to Internet when the WaveRunner is used with Internet Chameleon, a software product from NetManage, Inc.; this API also extends the interoperability of WaveRunner with other ISDN products, such as MAX Digital Access Server from Ascend Communications, Inc., which supports Point-to-Point Protocol (PPP)
- The following non-IBM programs are shipped with the WaveRunner Digital Modem:
- Fax application for Windows:
- WaveRunner, when used with this application program running under Windows 3.1, allows a personal computer to send or receive faxes, and customers can take immediate advantage of WaveRunner's fax capability
- Accelerator:
- A set of communication drivers for Microsoft Windows for rapid high-speed data transfers

Specifications
- Micro Channel or ISA card
- 16 MHz 386SX, or above, system processor
- 16-bit ISA or Micro Channel-compatible machines capable of supporting DMA bus master operation

- 4MB or more RAM, excluding cache, for Windows 3.1 environment
- 8MB or more RAM, excluding cache, for OS/2 environment
- At least 10MB HD space must be available to load WaveRunner installation, configuration, and diagnostics code
- VGA display monitor
- 3.5-inch, 1.44 MB floppy disk drive
- PS/2(R), PS/55, or Microsoft-compatible mouse
- A Network Terminator (NT1), such as the IBM 7845 NT Extended, for connecting the WaveRunner to an ISDN network
- Supports communications with the following modem types: Bell 103, Bell 212, V.22, V.22bis, V.32, and V.32bis
- Supported PSTN switch types: AT&T 5ESS with 5E7, 5E8, and 5E9 (Custom and National ISDN-1), Northern Telecom DMS-100 with BCS-33, BCS34, and BCS35; national ISDN-1-compatible switches, Euro-ISDN-, 1TR6-, and VN4-compatible switches
- Software requirements: Microsoft Windows 3.1 (Enhanced Mode) running under IBM DOS 5.0, or equivalent IBM OS/2 2.1, or later
- Limitations: Only one Mwave™-compatible adapter per system is supported
- A 10-foot cable, terminated with RJ45 plugs, is shipped with the product.
- Adapter diagnostics, designed to assist the customer in problem determination, are included in the WaveRunner diskettes
- Packaging: One box; the contents of the box for each of the WaveRunner Digital Modem part numbers listed below are one full-size 16-bit adapter, Micro Channel or ISA 3.5-inch diskettes, 10-foot communication cable, installation and testing instructions
- Warranty period. WaveRunner Digital Modem Adapter: one year

List Prices

IBM WaveRunner packaged with IBM 7845 NT1E	$845
IBM WaveRunner ISA	$545
IBM WaveRunner MicroChannel	$575
IBM WaveRunner PCMCIA	$595

- IBM 7845 NT-1E $350
  (Note: Even IBM discounts these prices.)

Comments

The WaveRunner is an impressive unit. Of course the software and support sound excellent, and they probably will be when all the features finally get delivered (IBM is notorious for putting all anticipated features in the information and spec sheets, and then delivering them incrementally over months or years). IBM does seem to be delivering its promised software upgrades nearly on time, however, with few serious complaints from customers on the news groups and listservs.

Keep in mind that the WaveRunner is an internal adapter card but not a network card. It's a serial device modem as far as your computer is concerned. There have been some setup problems noted on the *comp.dcom.isdn* news group, but the two IBM experts who answer all questions on the news group, and via telephone as well, have taken care of these problems quickly. One fellow is from the WaveRunner group and the other from the NT1E group at IBM, but they both appear strongly motivated to work issues out quickly and accurately.

For the price and the features, the WaveRunner/7845 NT1E combination is quite competitive in the combined ISDN/analog modem category. You should definitely check this one out when you're shopping for an ISDN modem. But be aware that it's an internal adapter card, and may therefore be more difficult to install than an external serial device.

Motorola TA210

The Motorola TA210 is an external ISDN terminal adapter with a built-in NT1 that acts as a "digital modem" when connected to your computer's RS232 serial port. The TA 210 is compatible with the National ISDN-1 standard and a wide variety of central office switches. The TA210 connects to the U interface basic rate ISDN, providing two 64-Kbps B-channels, one for circuit-switched data calls, the other for voice calls. The TA210 data port supports asynchronous data transfer rates up to 115.2 Kbps, allowing actual throughput of over 70 Kbps on any type of data. Synchronous data rates of up to 64 Kbps are also supported. The voice port functions as a standard telephone line, allowing you to connect phones, modems, or fax machines.

A built-in LocalMenu feature, accessed using an asynchronous VT100-compatible terminal, gives you the simplicity of menu-driven configuration. You can also use AT commands augmented with a set of Quick Setup factory defaults for faster application setups. If your PC is running Windows 3.1 or higher, the TA Configuration Manager utility provides on-screen instructions and context-sensitive help to guide you through configuring the TA210. The TA210 supports ISDN Solutions to ensure that you can order the ISDN services you need from your telephone company with a simple capability code.

Features

- Data rates up to 115.2 Kbps asynchronous; 1200 to 64,000 bps synchronous to accommodate a wide range of DTE devices
- Interfaces to ISDN basic rate interface (2B+D) providing access to two 64-Kbps B-channels, one for data calls and one for voice calls; access to a 16-Kbps D-channel for transferring signaling information
- Simultaneous voice and data operation lets you connect terminals, PCs or other data equipment and standard telephones to an ISDN line through a single U interface
- Digital Services Compatibility allows the TA210 to exchange data calls with Switched 56 and Datapath services
- Full POTS (plain old telephone service) support allows connection of existing telephones, modems, facsimile machines, and other devices that normally connect to a telephone line; supports REN up to 3.0B; provides true ringing voltages
- Multiple rate adaptation protocols CCITT V.120 and clear channel; CCITT V.120 gives you the capability to operate at lower DTE speeds
- Multiple signaling protocol compatibility with the following network switches: AT&T 5ESS, Northern Telecom DMS 100, North American National ISDN-1 (NI-1)
- ISDN Solutions compatibility to simplify ordering and configuring your ISDN service and equipment.
- AT command set compatible for use with a wide variety of asynchronous communications software packages
- V.25bis dialer for synchronous dialing applications
- LocalMenu menu-driven configuration

- Windows configuration program
- HyperACCESS data communications package is included
- Windows Plug and Play COM compatible

Specifications

- Synchronous data rates: 64,000, 56,000, 38,400, 19,200, 9600, 4800, 2400, 1200 bps
- Asynchronous data rates: 57,600, 38,400, 19,200, 9600, 4800, 2400, 1200, 300 bps
- Rate adaptation: CCITT V. 120; Clear Channel
- Signaling Protocol Compatibility: National ISDN-1 (NI-1); AT&T 5ESS central office switch; Northern Telecom DMS 100 central office switch
- Data terminal interface: ElA-232, CCITT V.24
- Line type: 2B1Q interface (ANSI T1.601, 1992)
- Size: width 6.4 inches; height 1.5 inches; depth 5.3 inches
- Power requirements: 115 VAC ±10%, 50/60 Hz
- Power consumption: 3 watts nominal
- Test features: Self test, 2B+D loopback, digital loopback, local terminal loopback

Comments

The Motorola TA210 is an impressive external serial ISDN modem with a strong list of features and software; however, its lack of a PPP protocol severely limits its use for connecting to many Internet service providers. If Motorola upgrades its software support to include PPP, the TA210 will be a serious contender in the digital modem category.

Motorola BitSURFR

BitSURFR from Motorola is a terminal adapter for high-speed, digital ISDN telephone service. Ideal for the small office, branch office, or work-at-home user, BitSURFR is an all-in-one combination of hardware and software ready to connect your PC, telephone, and fax machine to ISDN.

BitSURFR includes a built-in NT1 interface, so no other equipment is required for full ISDN service. And you can connect to Windows Plug and Play right out of the box. The Motorola BitSURFR is delivered complete with the ISDN Solutions ordering code that tells your telephone company exactly how to set up your ISDN line. BitSURFR's Windows Configuration Manager's simple configuration options quickly get you up and running.

Specifications

- Data rates: 300 to 115,200 bps asynchronous, 1200 to 64,000 bps synchronous, up to 128 Kbps with synchronous BONDing
- Standards: National ISDN-1 (signaling protocols), Northern Telecom DMS 100 central office switch, AT&T 5ESS central office switch, Bellcore SR-NWT-001953
- ISDN interface: U interface (2B1Q encoding per ANSI T1.601)
- Digital terminal interface: Standard EIA-232E, CCITT V.24 (for your computer)
- Phone interface: Standard RJ11
- Computer compatibility: PC with a 386 or 486 processor (or higher) and a 16550 or equivalent high-speed serial port
- Data compression: provided in HyperAccess bundled software
- B-channel protocols V.120, PPP (point-to-point protocol), BONDing
- Command set: Hayes; compatible AT commands
- Case dimensions: 6.4 in. wide, 5.3 in. deep, 1.5 in. high
- Weight: 10.5 oz.

Comments

The BitSURFR is Motorola's recent entry into the low-cost ISDN modem market. Little information is available about it. Motorola's history of solid products makes it worth watching and inquiring about when you get ready to make the jump into ISDN, if you want a digital modem T/A. If it lives up to its specs and is priced low enough, it should be quite a good buy.

Racal-Datacom ISDN BRI 2000 A-DAP-ter

The BRI 2000 is the first product in the Racal-Datacom family of Excalibur A-DAP-ter ISDN access products. The BRI 2000 provides an on-demand interface between your existing data terminal equipment (your computer) and ISDN Basic Rate service. The BRI 2000 operates at up to 64 Kbps per port and supports NI-1, AT&T, and NT switched services.

The BRI 2000 connects directly to the two-wire ISDN Basic Rate service through its "U" interface, eliminating the need for a Network Termination (NT1) converter box. The BRI 2000 provides a 32-character menu-driven front panel display that makes monitoring, configuring, and testing quick and easy. Clear English prompts guide you through each operating procedure.

Features

- 2-wire Basic Rate access to the ISDN (NT1 built-in)
- Dual DTE ports
- Network compatible with two- and four-wire switched 56 Kbps services
- Menu-driven front panel display
- CMS network management compatibility
- Password and CLID security
- Rate adaptation

Comments

Racal-Datacom's entry into the low-cost ISDN modem fray, the BRI 2000, looks good on paper as far as such information is available. Only time will tell whether it lives up to its specs and feature list. Here again, you'll want to check the newsgroups to see how its pioneering purchasers like it (and how well the company supports it).

ZyXEL Elite 2864I

Designed as a universal communication platform, the ZyXEL ISDN Elite 2864I modem gives you both the V.34 and ISDN worlds by allowing you to access ISDN's integrated services without sacrificing connectivity to today's more prevalent analog devices. This modem fully supports the National ISDN-1 standard, AT&T 5ESS, and Northern Telecom DMS-100.

The ISDN modem is compatible with PSTN and works with most communications software using AT commands. The ZyXEL Elite 2864I has a 4MB flash EPROM that allows you to conveniently download and program the modem with new firmware. The ISDN Elite 2864I modem has an analog adapter that recognizes standard DTMF tones and generates a standard ring signal to the connected device.

Features

- Compatible with NI-1, AT&T 5ESS, Northern Telecom DMS-100, 1TR6, and EDSS1 switches
- Rate adaptation ITU-T V.120 and V.110
- Ultrahigh-speed modem—up to 28.8 Kbps
- EIA-class 1, 2, and 2.0, ZyXEL fax AT commands
- ADPCM, voice digitization with speech compression
- Flash EPROM memory for easy firmware upgrades
- Optional 8MB DRAM for standalone fax receive/storage
- Parallel port for direct fax printing
- Combine two B Channels for 128 Kbps
- V.34
- ZyXEL 19.2 Kbps, 16.8 Kbps (proprietary)
- V.32bis/V.32
- V.26bis/V.22bis/V.22/V.23/V.21
- DTE serial interface to 460.8 Kbps
- V.17 G3 fax—14.4 Kbps, send and receive
- ISDN device with PSTN compatibility
- V.25bis and ZyXEL ISDN AT commands, CAPI 1.1a and 2.0
- Supports NetWare CAPI Manager
- Operates in DOS, Windows, OS/2, Macintosh, UNIX, Novell, Amiga, and IBM AS/400/RS6000 environments
- Compatible with a wide range of communications and fax software
- MNP 4/3 and V.42 error correction
- MNP 5 and V.42bis data compression

- Fast and robust ZyCellular protocol for demanding cellular operations
- High-speed serial/parallel DTE interface
- Routes incoming faxes directly to printer
- Automatic data/fax/voice call detection
- Computer telephony
- Fast retrain with auto fall-forward and fall-back
- Asynchronous/synchronous modes
- Call-back security with password protection
- Caller ID and distinctive ring
- Remote configuration capability
- Data and fax transmissions over cellular networks

Specifications

- Flow control: Software XON/XOFF or hardware CTS/RTS
- Configuration setting: Software programmable with nonvolatile memory for profile storage
- Diagnostics: Self test, analog loopback (with self test), digital loopback, and remote digital loopback (with self test)
- Line interface: RJ45, RJ11
- Call progress monitoring: dial tone, busy, and ring back detection
- ZFAX data/fax/voice software included
- Five-year warranty on parts and service
- Knowledgeable technical support
- 24-hour BBS technical support line

Comments

As with some of the previously discussed ISDN modems, the ZyXEL ISDN Elite 2864I offers an impressive list of features and specifications. Especially noteworthy are the 28.8 Kbps analog modem, ADPCM, voice digitization with speech compression, flash EPROM memory, optional 8MB DRAM for standalone fax receive/storage, and parallel port for direct fax printing. Again, if this product lives up to its promises, it may be competitive in the ISDN modem market. This will depend on pricing (which we'd expect to be higher, because of the number and kinds of features), and how well the unit works for its early adopters.

Network Termination Devices (NT1s)

This section describes standalone NT1s. The Network Termination 1000 device (NT1) provides the connection between the telephone company ISDN jack in your wall and the ISDN Terminal Adapter (T/A) device, which in turn connects to your computer. As you may have already seen in the previous sections, many T/As have built-in NT1 devices. This is popular in the United States, where the consumer must usually purchase an NT1. In Europe, the NT1 is considered part of the telephone company's equipment; therefore, all European T/As are sold without NT1s.

The following listing and brief discussion of NT1s is provided primarily for those of you who want an external device for your ISDN T/A device(s), and possibly for your analog phones (that is, if the NT1 has a POTS jack). External NT1s may be powered by UPSs with battery backups to keep your telephone service up and running for up to 8 hours during AC power outages. They also allow you to keep your phones and fax equipment turned on continuously without keeping your computer on.

As this section was written, the list prices for NT1s ranged between $200 and $350, but some sales prices were as low as $175. These prices should continue to fall in the U.S., followed by increased usage by individuals installing ISDN lines in their homes and businesses.

Since the NT1 has been exclusively a business item until recently, little sales literature or other consumer information is available for most of the models listed below. They have been treated more like "black-box" devices than feature-filled consumer items by virtually every manufacturer (except for IBM; IBM has added a considerable number of features to their 7845 NT1E for those of you who want more than a plain NT1).

Your best bet will be to call around immediately before you purchase an NT1 to make sure you're getting the latest model with the desired features at the best price. A list of NT1 manufacturers and vendors is included in an appendix to this book.

ADTRAN—ISDN Network Termination (NT1 ACE)

ADTRAN NT1s support the ANSI 2B1Q line coding and multiple switch vendors including AT&T, NEC, Northern Telecom, and Siemens.

NT1 ACE

Small, plastic, standalone unit appropriate for desktop use. Can be powered by external power supplies such as the ISDN PS2 or the NT1 power supply kit.

ISDN PS2 Power Supply

Designed to complement the NT1 ACE, the PS2 provides up to 12 watts to the NT1 ACE and terminal equipment. It provides 9.5 seconds of reserve power to terminal equipment to bridge AC power interruptions.

NT1 Power Supply Kit

The power supply kit is a low-cost alternative power supply for the NT1 ACE. The kit includes a basic unregulated wall-mount power supply (@ 10 watts), a T adapter, and an RJ11 cable that links power to the NT1.

Alpha Telecom, Inc. UT620 ISDN NT1 Network Termination Device

The UT620 is an affordable ($225) and compact ISDN NT1 device that provides both an ISDN telephone and a terminal adapter (TA) port for access to the ISDN network. The UT620 NT1 device is designed to conform to the ANSI T1.601, 1992; T1.605, 1992; ITU/CCITT I.430 standards; and industry standard 2B1Q line code for the U-interface. The UT620 is easy to install as a desktop or wall-mounted unit for the office environment.

Features/Specifications

- Network U-interface
- Line: two-wire, full duplex
- Data rate: 144 Kbps available to customer
- Line code: 2B1Q per T1.601-1992
- O/P amplitude: 2.5 V, zero to peak

- Connector: one RJ45 or one RJ11
- Automatic ANSI maintenance functions
- Terminal S/T-Interface
- Line: four-wire, full duplex
- Line code: AMI, 100% duty cycle
- Conforms to ANSI T1.605, 1991-TX Source Impedance, RX Impedance, Receive Sensitivity
- Connectors: two RJ45
- Supports point-to-point and point-to-multipoint
- DIP switch
- S/T-interface termination resistance
- S/T bus timing mode
- PS2 power selection
- Standalone desktop
- Wall-mount
- Rack-mount
- 1.3"×3.93"×5.7"
- 0.97 pounds
- Two-year warranty

AT&T NT1 L-230 Network Terminating Unit

AT&T has added another network terminating unit to its NT1 product line. Called the in-line NT1 L-230, the new unit is smaller and lower in cost than previous models. The shirt-pocket-size unit can be attached under a desk to save space. The new terminating unit features LED status indicators and provides multipoint data connections. It has a manufacturer's suggested retail price of $230.

IBM 7845 Network Terminator Extended (NT1)

The IBM 7845 ISDN Network Terminator Extended is a programmable, standalone device that allows a personal computer or a workstation equipped for basic-rate ISDN and standard analog telephone lines to share a single digital phone line into the home or office. The Network Terminator (NT) Extended provides the connection between the telephone company jack and the ISDN TA. It will also connect standard analog telephone equipment to the ISDN network over the same ISDN basic-rate service. This allows customers to replace their existing analog telephone service with one of the ISDN B channels while continuing to use their existing phone equipment.

The analog telephone function of the NT Extended offers several key custom calling features similar to those offered by the telephone companies as options with regular telephone service.

Features

- Speed dialing
- Redial of last number dialed
- Repetitive redialing of the last busy number
- Return of the last incoming call
- Call hold, call retrieve
- Call waiting
- Call blocking
- Three- or six-way conferencing

A rechargeable battery, included with the NT Extended, provides backup power to the analog phone service should a temporary power outage occur

For more information on the IBM 7845 NT Extended, refer to Hardware Announcement 194-252, dated July 26, 1994.

Motorola NT1D

The NT1D is an NT1 unit providing interface between ISDN network and ISDN terminal equipment in both point-to- point and multipoint configurations.

Specifications

- Line rate: 192 Kbps

- Line interface: 8-pin modular (RJ45); U interface, ANSI T1.601-1992.; maximum length: 18,000 feet on 26 gauge wire.
- Customer interface: 8-pin modular (RJ45). S/T interface, ANSI T1.601-1991.
- Size: 6 in. (W) x 5 in. (H) x 1.5 in. (D); wt.: 2.5 lb.
- Power: 120 VAC; 5 Watts.
- Supplied: Wall-mount transformer; RJ11C to RJ45 cable (6 ft.) for U interface.
- $225.00 list
- Order #6457503600010 Network Termination 1 Device

Summary

The list of ISDN modems is impressive, but it can also be confusing. Their veritable plethora of features—that most of us will never use—look good in theory, but will they help us to "get the job done?" Some of them undoubtedly will help, but others may not.

The bottom line on these devices it to check your requirements carefully against the list of features and the price of the ISDN modems. If you're not careful, you'll spend more for a slower modem device with a lot of features you don't really need, than you would if you purchased another type of ISDN T/A. But if you want simplicity and ease of installation, you're looking in the right place when your considering an external ISDN modem with a built-in NT1.

The list of NT1s is short but contains known producers of quality products. Check the extra features and give one a try if you're in the market for a standalone NT1. The IBM 7845 appears to have the most features, but may also have the highest price. However, you can probably find it for less than $300, and that makes it only about $100 more than the lowest-priced NT1.

A final note: You may be happier with your entire ISDN system if you get an external NT1 even if you get an internal adapter card, since you won't have to keep your computer running to use your ISDN phone number, either via an ISDN phone or your analog phone plugged into the POTS jack on the NT1.

chapter 8

ETHERNET ISDN BRIDGES AND ROUTERS

WHAT ARE ETHERNET ISDN BRIDGES AND ROUTERS?

Bridges and routers are both network devices that pass LAN packets of information from one network node or machine to another. They differ in that bridges pass along LAN packets without looking at the network addresses in the packets, whereas routers pass the packets to the proper network address (or to other routers it believes are closer to that address). In the simplest sense, bridges merely link two network devices, but routers distribute the information they receive more intelligently.

Routing functionality is key to advanced functions that are frequently important to larger business users. Routing is the basis for SNMP, security protocols, call management, and TCP/IP traffic management. These functions are necessary for telecommuting, Internet, and business-to-business networking applications for multiple users at the same location.

Both bridges and routers connect to an existing network, most commonly Ethernet (but support many other topologies, including token ring, ARCnet andFDDI). This section provides information on several currently available Ethernet bridges and routers that include ISDN capabilities. Unless you already have an Ethernet card in your computer, or an Ethernet network in your immediate neighborhood, you probably won't be interested in these devices. They are higher-priced and more complex than either ISDN adapter cards or ISDN modems.

ISDN-capable Ethernet bridges and routers, while the most versatile and powerful of the ISDN devices discussed in this book, are really designed more for high-volume business use than for individuals seeking to connect to the Internet via an ISDN line, or to make a few phone calls with an analog phone plugged into the POTS jack of the ISDN device.

We have included them for any of you who may want to expand your system in the future, or who just want to know what the next step up looks like. Since these devices are more complex, we have included only a brief description for each of the more popular and lower-priced models. For pointers to the WWW pages for each of the manufacturers (and much more detailed information on each device), we suggest you check out Dan Kegel's ISDN WWW site at `http://alumni.caltech.edu/~dank/isdn/` or that you contact these manufacturers directly, using the information in the book's appendices. The following information is listed by manufacturer in alphabetical order. Please note also that the lowest-priced ISDN bridge is about $1000.

ISDN Bridges and Routers with Ethernet Interfaces

ACC Nile—The Branch Office Access Router

The Nile is a remote multiprotocol bridge/ router designed to minimize network operating costs, while maximizing connectivity options, performance, and flexibility. A true access router, Nile is ideal for remote branch offices, working together with other ACC products or third-party routers to provide an enterprise network solution. Nile supports all popular wide-area protocols, including frame relay, X.25, SMDS, dedicated lines and ISDN. Nile also supports ACC's exclusive Bandwidth Optimization features, which dramatically reduce the lifetime operating costs of your network.

Nile is available in Ethernet or token-ring LAN models, with either one or two WAN interfaces. All WAN interfaces are field-installable, with installation so simple that interfaces can be added or changed without opening the enclosure.

Network-level routing of today's most popular protocols is another way Nile delivers superior connectivity. Routing protocols can be activated individually, so Nile can adapt to your changing network needs. Since Nile software supports multi-protocol routing and simultaneous bridging, you can bridge where protocol transparency is needed, route where path control is necessary, or combine them for increased benefits.

At Nile's heart is a 25 MHz 68020 processor. This processor, in combination with a high performance memory architecture, provides the performance needed in today's demanding networks. Nile supports combined link speeds between 9.6 Kbps and 2.048 Mbps.

Nile's memory can be expanded from 2 to 8 MB to handle networks with large routing databases.

Ascend Pipeline 50 HX

The Ascend Communications, Inc. Pipeline 50 HX is a full-function ISDN bridge and IP router. The Pipeline 50 HX connects single users at home and in small offices to backbone networks and the Internet over inexpensive, high-speed ISDN lines.

Ascend's Pipeline 50 HX connects Ethernet through ISDN Basic Rate Interface lines, with throughput of up to 512 Kbps on dial-up connections. Pipeline 50 HX integrates full routing a learning bridge, inverse multiplexing, data compression, and an ISDN TA with an optional internal NT1 in a compact package scarcely larger than a videotape.

The Pipeline 50 HX is the newest addition to the Pipeline Access Server family. With dimensions of only 8.2" x 6" x1.2", the Pipeline 50 HX provides an Ethernet interface, and an ISDN BRI S/T interface or an ISDN BRI U interface with an internal NT1. The Pipeline 50 HX includes standards-based bridging and IP routing, ISDN dial-on-demand, inverse multiplexing of the two ISDN B channels, 4:1 data compression, SNMP remote management, and extensive password, challenge, and token-based security features.

Routing functionality is key to advanced functions that are important to corporate buyers. Routing is the basis for SNMP, Security protocols, call management, TCP/IP traffic management, etc. These functions are necessary for telecommuting, Internet, and business-to-business networking applications. For example, routing is necessary for the Pipeline 50 HX to do real-time traffic analysis to determine if the line has no activity and should be shut down.

When new activity is initiated the line is reestablished, thereby saving customers hundreds of dollars a year in unused connect time. Another unique feature is upgradability to multi-user routing with a remote software upgrade, providing small business an economical way to upgrade as they grow.

The Pipeline family provides the other necessary products for all corporate remote requirements: home, branch, regional, and central site solutions. Pipeline MAX Integrated Access Servers reside on the enterprise network and provide access for up to 96 simultaneous remote connections, using either ISDN or analog modems. Pipeline 400 Access Servers can reside on the enterprise network for smaller concentration tasks, or can provide higher-speed connections for remote workgroups, using ISDN BRI, ISDN PRI, T1, or Switched 56 network services. The Pipeline 50 Workgroup Server is for high-speed branch office communication using ISDN BRI or Switched 56 network services.

Pipeline 50 HX—S/T Interface purchase price $ 995
Pipeline 50 HX—U Interface Purchase Price $ 1,195

Bay Networks Access Node/Access Node Hub Router

The Wellfleet Access Node (AN) router family from Bay Networks cost-effectively connects small remote offices and ensures network availability while minimizing network operating costs. Typical connectivity requirements of remote offices are supported by the AN's LAN interfaces (a single Ethernet, a single token-ring, or both an Ethernet and a token-ring), and serial interfaces (two synchronous interfaces or one synchronous and one ISDN BRI interface).

Ethernet ANs are also available in fully managed 8-port or 12-port Access Node Hub (ANH) configurations, reducing equipment and management complexity. The 8-port ANH also features one AUI port, and both ANH configurations support two serial interfaces. In all configurations, the two serial interfaces provide remote office network design flexibility. For mission-critical applications, they facilitate dial back-up and bandwidth-on-demand support. Also, dial-on-demand functionality enables ANs to extend network availability on an as-needed basis to small remote sites, minimizing WAN service costs. Optionally, one synchronous interface may be used for SDLC, allowing the consolidation of IBM SNA and multi-protocol LAN traffic over one WAN link to the internetwork backbone.

The AN easily integrates with the enterprise internetwork, supporting all major LAN and WAN protocols. The Motorola 68360 processor, used in the AN's highly integrated design, easily maintains high forwarding and filtering rates, regardless of the number of protocols and network interfaces used—even when processing SNMP management inquiries. Multiline circuits, uniform traffic filters, traffic prioritization, and data compression optimize bandwidth use and maximize traffic control. Additionally, EZ-Install, EZ-Configure, and EZ-Update simplify installation, configuration, and software maintenance. The AN is easily configured and monitored via Bay Networks Optivity/Internetwork, which combines three applications—Site Manager, RouterMan, and PathMan—to form a seamless integrated router management package for Wellfleet routers.

The AN complements the Wellfleet Access Stack Node (ASN), Backbone Link Node (BLN), and Backbone Concentrator Node (BCN) to satisfy all connectivity, performance, and availability requirements ranging from cost-effective workgroup or remote site access to high-performance highly available network hubs. The Wellfleet product family connects up to 52 network interfaces and delivers aggregate system forwarding performance up to 760,000 bps.

The AN's ANH configurations integrate the function of both a router and a hub in one platform, simplifying equipment and management requirements. The AN's EZ-Install feature reduces installation time and expense by enabling the AN to get its IP address and configuration from a central site. Remote site software maintenance time and expense are also minimized by the AN's EZ-Update feature, which allows software updates to be downloaded from a central site.

Site Manager's EZ-Configure feature enables configuration file templates to be copied or modified for use at other sites. Site Manager also supports a Dynamic Software Builder and Loader function that preserves flash memory and DRAM space. Data compression support also minimizes costs by enabling lower cost links to achieve the throughput rates of higher-cost links. SNMP-based management applications monitor and control nodes with a single graphical user interface. Small remote offices can also connect to the internetwork on an as-needed basis via the AN's dial-on-demand support.

Cisco 1003 ISDN Router

The new Cisco 1003 ISDN router connects small, remote sites with Ethernet local-area networks (LANs) to wide-area networks (WANs) using ISDN at speeds up to 128 Kbps; with 4:1 data compression, raw throughput speeds of 512 Kbps are possible. The Cisco 1003 has a built-in ISDN Basic Rate Interface (BRI) port, a 10BaseT Ethernet port, a console port, and an external Personal Computer Memory Card International Association (PCMCIA) slot for a Flash ROM card. This "plug-and-play" product is designed to be installed easily by nontechnical personnel at remote sites.

The Cisco 1003 supports two software feature sets, based on the Cisco Internetwork Operating System (Cisco IOS). One set includes IP routing and transparent bridging; the other contains IP and AppleTalk routing plus transparent bridging. Both software sets support Point-to-Point Protocol (PPP), compression, dial-on-demand routing (DDR), and a host of other powerful features for optimizing WAN bandwidth and costs.

Features

- One Ethernet 10BaseT port with RJ45 connector
- One ISDN BRI port with RJ45 connector
- One console port with RJ45 connector
- One external PCMCIA slot for Flash ROM card
- Optional Flash ROM card stores bootable software image and allows software updates over the WAN or LAN connections
- LAN-to-LAN dial-on-demand routing (DDR) over ISDN lines
- IP, IPX, and AppleTalk routing
- IPX spoofing
- PPP compression and bandwidth on demand with load balancing for throughput up to 512 Kbps
- Standards-based PPP, HDLC and LAPB

Combinet EVERYWARE CB-160 Access Unit 1 BRI—Bridge

Combinet's EVERYWARE CB-160 Access Unit provides a single user access to an enterprise site. The CB-160 is compatible with any computer with an Ethernet interface and allows access to all applications and data resident on an enterprise network.

The CB-160 provides individual users with access to the enterprise LAN using ISDN BRI telephone lines. The CB-160 sends packets over two 64 Kbps digital channels simultaneously, for a total data rate of 128 Kbps in each direction.

Features

- Built-in NT1
- Data compression for high performance
- Supports authentication and callback security
- Flash memory for easy software upgrades
- Nonvolatile memory to save configuration data
- Informative front panel status indicators
- Easy-to-use configuration menu
- Remote configuration over Ethernet or ISDN
- On-demand or manual call control LAN interface as 10BaseT hub

Combinet EVERYWARE CB-200 Access Unit 1 BRI—Bridge Multiuser

Combinet's EVERYWARE CB-200 Access Unit provides remote offices and workgroups access to an enterprise site. The CB-200 can call into an enterprise LAN and access all applications and data resident on the LAN.

The CB-200 connects remote users to the enterprise LAN over Integrated Services Digital Network Basic Rate Interface telephone lines. The CB-200 sends packets over two 64 Kbps digital channels simultaneously for a total data rate of 128 Kbps in each direction (before data compression). The CB-200 can call to a remote CB-200/300/400, a remote CB-600 over 56 Kbps lines, or a remote CB-900 Access Server over ISDN PRI lines.

Features

- Data compression for high performance
- Compatible with EVERYWARE single-user products
- Supports authentication and callback security
- Flash memory for easy software upgrades
- Nonvolatile memory to save configuration data
- On-demand or manual call control
- Informative front panel status indicators
- Easy-to-use configuration menu
- Remote configuration over Ethernet or ISDN

- 10BaseT and 10Base2 Ethernet ports
- 10BaseT port configurable as hub

Gandalf LANLine 5242i Telecommuter Bridge

Gandalf's LANLine 5242i Telecommuter Bridge provides 128 Kbps ISDN and analog voice interfaces, plus up to 4- to-1 data compression for high-performance LAN connectivity for homes, small offices, and telecommuters. Providing on-demand bandwidth for remote users, the 5242i gives a virtual presence on the central LAN. Connections are made in a few seconds, simply by requesting a remote IP or IPX address. The 5242i also supports IPX spoofing and is compatible with Gandalf's 5240i and 5225i central site intelligent bridges, as well as the Gandalf Xpressway, the highest-capacity Central Site communications server on the market.

One of the unique features of the 5242i is the ability to handle both voice and data calls simultaneously. A user may be busy transferring data over both B channels and either receive or originate a voice call, using the POTS (plain old telephone service) interface provided on the 5242, without stopping the data transmission. If someone calls the user while he or she is busy transferring data, the 5242i is notified of an incoming telephone call via the 16 Kbps D channel, when the 5242i detects the incoming call, it seamlessly brings down the second 64 Kbps B, channel allowing the voice call to come in. Likewise, when the user picks up the phone to make a call out, the 5242i reserves the second B channel for voice use. Data communications continue uninterrupted on the primary data channel.

The Model 5242i, data only, is	$920
The Model 5242i, voice/data, is	$1160

Features

- ISDN BRI(U), analog voice, 10BaseT
- Address table supports 10 remote users
- Touchtone telephone configuration
- SNMP MIB II management support
- Flash memory for easy firmware upgrades
- High-speed data compression

- Connect/disconnect on LAN activity
- Overflow to second B channel upon primary B channel congestion
- Store 10 destination numbers
- Voice and data negotiation for use of one B channel for data access while the other provides voice service

Lightning MULTICOM-IPB TCP/IP Router-Bridge—
1 BRI + 1 WAN—IP Router; Bridge

The LIGHTNING MultiCom IPB Router-Bridge platform delivers wide-area connectivity for the widely used Ethernet environment. The MultiCom IPB Router-Bridge is a low-cost solution for linking networks, such as those found on the UNIX, PC and Macintosh workstations. This platform includes three ports: one Ethernet, one ISDN up to 128 Kbps, and a serial link for a leased line. Its excellent performance characteristics using load-controlled connection management satisfy extensive communications needs, even without leased lines. Its native simplicity makes it customer-installable. All management functions can be done remotely, including software upgrades. The platform is compatible with standards-compliant products from third-party vendors.

The MultiCom IPB Router-Bridge operates at the data link level, filtering and forwarding the traffic according to Ethernet addresses. The operation of the MultiCom IPB Router-Bridge is transparent to all higher-level protocols, including NetBIOS, NetBEUI, Novell, AppleTalk, OSI, XNS, TCP/IP, and all the others. It is possible to selectively bridge specific protocols concurrently.

The MultiCom IPB Router-Bridge is easy to install and operate. It can be configured through the network from a remote management station or locally from an ASCII terminal attached to the serial console port. Clearly designed status lights provide continuous local display of Ethernet activity, serial link activity and ISDN status and activity.

Each MultiCom IPB Router-Bridge has a remote phone number authorization list, which can be configured, and an authentication procedure, which guarantees secure access only by authorized remote sites. The MultiCom IPB Router-Bridge uses the calling number authentication feature of ISDN and does not respond to calls from unauthorized numbers. The ISDN network's closed user groups feature is also supported. The access by the network manager to the configuration features of the MultiCom IPB

Router-Bridge is protected by an authentication procedure, preventing unauthorized users from reconfiguring the MultiCom IPB Router-Bridge.

The MultiCom IPB Router-Bridge can be configured on the fly without shutting down the power or the connections, allowing for zero downtime. Link Cost Control Both (LCCB) B-channels of the ISDN connection can be simultaneously used at full speed, doubling the available ISDN throughput. The links are automatically opened and closed during operation, using a load and cost-

A built-in Flash-EPROM electronic memory allows for easy remote software upgrades without physically access to the router. This memory also permanently stores the MultiCom IPB Router-Bridge configuration. The absence of moving parts makes the MultiCom IPB Router-Bridge a sturdy platform able to operate in dusty and hostile environments. Its low power consumption also makes it well-suited for emergency supply powered critical applications.

Network Express ISDN Router—20 BRI + 2 PRI—IP Router

The Network Express NE ISDN Router offers simple, flexible and cost-efficient methods to interconnect local area networks. It integrates switched digital services and inverse multiplexing allowing multiple LANs to simultaneously interconnect via switched digital services such as ISDN, Switched T1, and Switched 56. It provides high-speed bandwidth on demand to single or multiple locations simply by initiating or terminating telephone calls.

The NE ISDN Router routes packets received from the LAN to their correct destination. The NE ISDN Router integrates with other bridge/routers in existing networks to take advantage of switched digital services and can provide automatic backup and bandwidth peak overflow capacity for dedicated networks. Furthermore, remote sites that cannot justify the expense of a dedicated leased line can use the NE ISDN Router to provide high-bandwidth connectivity to an existing network.

The network's existing network management system (SNMP) can be used for monitoring and configuring the NE ISDN Router. The controllers can be configured from a central control site without disturbing the network operation.

Features

- Connects remote offices easily and economically to the corporate backbone
- Inverse multiplexing aggregates multiple channels to accommodate high speed data traffic
- Uses both dedicated and switched services to build a fault-tolerant, economical network that can adjust to peak traffic loads
- Dial-on-Demand muting to connect LANs through switched digital public and private networks
- Allows simple maintenance and field upgrades through remote management
- Provides a high performance industry standard router that supports the multiprotocol needs of the LAN outer network
- Supports Ethernet (802.3) LAN interfaces
- Integrates with other bridge/routers on enterprise LAN
- Bandwidth is automatically adjusted on demand thereby optimizing network price performance
- Provides simultaneous multiple connections to different locations within the wide area network
- Integrates T1, fractional T1, and other dedicated lines to switched connections which optimizes monthly line costs; provides automatic backup and accommodates peak load demands
- Supports domestic and international switched services such as ISDN Basic Rate and Primary Rate and Switched 56 services

Telebit NetBlazer LS ISDN and LS 2-PT—1 BRI—Routers

The LS ISDN provides LocalTalk synchronous/asynchronous ports, LAN protocol support, ISDN, frame relay, and 4:1 compression on sync and ISDN ports. It also provides higher bandwidth than TAs attached to the NetBlazer by offloading compression processing to the LS.

The LS 2-PT provides dual sync/async ports to leverage your investment in modems, ISDN TAs, or CSU/DSUs. Or, choose the LS ISDN, which also includes a single ISDN BRI port. You can connect the ISDN port to a regional or central site NetBlazer that contains a dual BRI card (NI2B).

All LS models include full router, terminal server, and modem pooling capabilities and support remote control or remote node access capability. The flexibility of the LS family of products allows you to invest only in what you need, and nothing more.

With the NetBlazer LS ISDN and LS 2-PT, your small branch offices and remote sites can connect not only to the corporate LAN, but to other remote office LANs as well. The LS connects over digital dial-up lines provided by the Public Switched Telephone Network (PSTN).

As a terminal server, the LS provides terminal sessions with network hosts to users who call in, locally or remotely. The LS can provide services to dumb terminals or PCs/workstations that perform terminal emulation. The terminals can attach directly to the LS's asynchronous ports, or they can dial in remotely through modems.

As modem pool servers, NetBlazers such as the LS ISDN and LS 2-PT allow users access to a common pool of modems for making outbound calls. Any user on the branch-office LAN can access services on the Internet, bulletin boards, and other value-added networks, such as Dialog, CompuServe, and Lexus/Nexus, without the need for a modem on every desk.

The NetBlazer LS makes remote servicing and troubleshooting easy. With the LS, network administrators can inexpensively access the remote network. The LS provides the ability to run SNMP over dial-up links. A network administrator can use SNMP to access any device on the network and run diagnostics, view network statistics, and configure workstations and routers as if connected locally. To extend the reach of your remote workstation to any IP, IPX, or AppleTalk LAN in the world, simply connect the LS ISDN (with NT-1) to your ISDN jack, or use the sync/async port on your LS ISDN or LS 2-PT to connect via modem or CSU/DSU.

SUMMARY

If you already have an Ethernet card in your computer and you understand your network reasonably well, one of the ISDN bridges and routers listed above will undoubtedly give you the best ISDN connection possible. Keep in mind that each of these products has a list of features and specifications that are too complex for all but network award persons to fathom. The best advice to you is to do-it-yourself only if you're highly knowledgeable about your network hardware and software, or if you are working from a remote location for a business that has a knowledgeable network administrator who will either physically help you at your location or at least remotely configure your bridge or router when you get it attached to your network card.

Check Dan Kegel's ISDN WWW site at `http://alumni.caltech.edu/~dank/isdn/` for the WWW pages of each manufacturer or contact the manufacturers directly via the information in the appendices for the latest products and prices before you order. Remember that these devices have been made for the business market and are primarily marketed through the normal business routes; therefore, you will find few of these devices advertised in the personal computer magazines.

chapter 9

ISDN Software

AN OVERVIEW OF ISDN SOFTWARE REQUIREMENTS

All of the ISDN hardware devices discussed in this book come with their own installation and configuration software or a physical method of setting them via DIP switches or telephone keypad commands. The software discussed in this chapter includes those additional programs you may need to get your chosen ISDN device to work with your computer system (usually network driver software), to help you make ISDN voice phone calls (telephony software), and to connect to your Internet service provider (ISP). The amount of additional software you will need to implement your ISDN system depends on the type of hardware you decide upon.

Internet Protocol and Connection Software

The primary Internet protocol software you will need for your PC running DOS or Windows is the TCP/IP stack. The core TCP/IP protocols are Transmission Control Protocol (TCP), Internet Protocol (IP), User Datagram Protocol (UDP), Address Resolution Protocol (ARP), and Internet Control Message Protocol (ICMP). This suite of Internet protocols provides a set of standards for how your computer will communicate with the Internet. Various TCP/IP stacks are available as shareware, as freeware, or included in commercial Internet packages. Your ISDN device will need to be compatible with your TCP/IP stack to connect with the Internet.

The most common of these are Trumpet, MS TCP/IP-32, and NetManage's TCP/IP. All of these run under Windows 3.1, WfW (Windows for Workgroups 3.11), or Windows NT 3.5. The Windows Sockets application programming interface (API) was developed to provide a standard application interface to different vendors' protocol implementations. The official name for the dynamic-link library (DLL) used for Windows Sockets 1.1 support is WINSOCK.DLL. All of the TCP/IP stacks mentioned above support the Windows Sockets 1.1 API, allowing common Windows Internet applications, such as Netscape, Mosaic, Chameleon, FreeAgent, or Eudora to run.

No matter which type of ISDN device you choose, you will also probably need PPP software to create a dial-in connection to an ISDN ISP. Although there is little difference between SLIP and PPP from the user's standpoint, it seems that the ISDN hardware manufacturers have provided only PPP software, both for the remote (user) end and the server (ISP) end. Most, but not all, of the ISDN hardware products discussed in this book have PPP software included in the package, or the product is PPP-compatible. Those that don't undoubtedly will have it available soon or they will be out of the market. Most are going toward MPP (Multiple Point-to-Point Protocol) in the future, so check for this feature in the specs of your chosen ISDN device.

ISDN Modem Software

If you choose an ISDN modem to replace your current analog modem for your Internet connection, you shouldn't need to change any of your software. ISDN modems respond to the Hayes AT command set, so you should be up and running very quickly.

ISDN PC Adapter Card Software

Choosing a PC ISDN adapter card will create the need for network driver software to link the card with your computer's operating system and interface program (Windows, for example). This gets you into the various layers of operating system programs, interface programs, network drivers, and LAN packet drivers. Some of the drivers you will see on PCs are NDIS,

ODI, TCP/IP, NetBEUI, and NetBIOS. Your chosen ISDN hardware manufacturer should provide the appropriate drivers if the device needs drivers other than those that are normally included with DOS, Windows, Windows95, or OS/2 and WARP.

You may still need to obtain a different TCP/IP "stack" if your system doesn't like the one you're currently using. For example, if you're currently using Trumpet Winsock with WfW and SLIP to connect to the Internet, you may need to switch to Microsoft's TCP/IP-32 instead of Trumpet, since the NDIS drivers may not be able to connect between Trumpet and the ISDN card. This isn't all that painful, since MS TCP/IP-32 is free from Microsoft. But all of this acronymspeak does give you an idea of how complicated the software is for an ISDN network adapter card.

ISDN Ethernet Bridge or Router Software

The final level is the ISDN bridge or router that connects to your existing Ethernet plug, assuming you have an Ethernet card in your computer with a network up and running. You will have driver questions to answer with this type of device that are similar to questions presented for network adapter cards. However, since you're not actually installing another network card, you shouldn't have to install any additional network driver software to get your computer to communicate with the bridge/router.

Still, this isn't as easy to do as it might look here, since you will have to reconfigure your network to recognize the additional device. If you're on a Novell or other "serious" network, you'll need the help of your network administrator to do this properly, without causing trouble for something else on the network. You will also have to install TCP/IP software to let your ISDN bridge/router dial and connect to your ISP.

Shareware and Freeware

The few current shareware programs available are primarily TCP/IP packet drivers and router software. Much of it is being produced by people in Germany for the European MS-DOS/MS-Windows and UNIX markets. Some has been tested on products available in the U.S. These packages are discussed later in this chapter.

Most of the shareware and freeware Internet application software has a built-in PPP client. These include Trumpet Winsock and MS TCP/IP-32.

You may need what is called a "shim" to put between Trumpet and your device's internal software to initiate the ISDN connection. Shim software is usually freeware from someone who has figured out the solution and is willing to give it away. Of course it isn't "supported," so you're on your own using it. Shims generally slow down the transmission and aren't the optimum way to go. But sometimes a shim is the only way to get things working.

Commercial Software

Other than the software provided with your ISDN device, most PC-based ISDN systems will only need a TCP/IP stack to connect to an ISDN ISP. Microsoft is giving their MS TCP/IP-32 away free for WfW 3.11 users. Windows95 comes with its own ISDN software and drivers. OS/2 Warp also provides its own TCP/IP stack and Internet software. All of these have adequate documentation for installation and start-up. It is helpful if you can get your ISP to help you with some of the settings that may be specific for their ISDN server equipment (generally an ISDN router).

ISDN Shareware/Free Software

MS-DOS/MS-Windows

ISDI: An MS-DOS NDIS Driver for ISDN Access

ISDI is a (real mode) NDIS-MAC driver for IP-Routing or remote Ethernet bridging over ISDN. ISDI communicates with the ISDN card using the ISDN API 1.1 specification (a standard defined by German ISDN card manufacturers and the German Telekom). Because of this, ISDI is completely hardware-independent and has successfully been tested with many active or passive ISDN cards.

ISDI was developed for Internet access over ISDN from WfW 3.11 and MS TCP/IP-32. ISDI has successfully been tested with Win95 beta releases. ISDI is also known to work with other NDIS-based TCP/IP packages for DOS and Windows, e.g., Chameleon.

ISDI was written for use with ISDN BRI PC cards and has been tested with the Teles.S0, one of the cheapest ISDN cards in Germany, as well as

NCP cards, AVM A1 & B1, Creatix S0/16, Diehl Diva, SCOM & S0Tec, Loewe ISCOM C100, MIRO, mbp Solis, NCP P16 & A, and Dr. Neuhaus NICCY 1000 PC.

ISDI supports the LAPB, Frame-Relay, PPP, SLIP, and Cisco-HDLC protocols for communication with ISDN routers or servers. ISDI can communicate with the following commercial systems: Ascend Routers, Biodata ISDN Router, Cisco Routers, Conet S2M Router, INS/CLS Banzai ISDN Router, netCS ISDN Router, RzK SLIP Bridge, SGI Indy ISDN 1.0, Spyder Routers, SunLink ISDN 1.0, and SunLink ISDN 1.0.2.

Some protocols conserve the protocol type over point-to-point lines (multi-LAPB, Frame-Relay, Cisco-HDLC). These protocols are able to do multiprotocol routing. For PPP only IP support is implemented at the network configuration layer.

The current version of ISDI supports two independent active connections at a time. Alternatively, a connection can use both B-channels for load-sharing. ISDI can be loaded more than once, if more than two simultaneous connections to different sites are desired. Load-sharing can be configured as static or dynamic (bandwidth on demand). Dynamic load-sharing can be used concurrently with a second independent connection. Load-sharing over two channels is implemented using simple round-robin scheduling, because IP doesn't require the original packet sequence.

ISDI shareware by Herbert Hanewinkel is available from ftp.biochem.mpg.de in /pc/isdn.

ISPA—MS-DOS Packet-Driver for TCP/IP over ISDN

ISPA is an Ethernet-type (class=1) packet-driver for IP routing or remote Ethernet bridging over ISDN. ISPA communicates with the ISDN card using the ISDN API 1.1 specification (a standard defined by German ISDN card manufacturers and the German Telekom). Because of this, ISPA is completely hardware-independent and has successfully been tested with many active or passive ISDN cards.

ISPA was initially developed for use with PCROUTE as a cheap Ethernet-ISDN router. However, it is used more and more to connect a standalone system to the Internet using ISDN. ISPA has been successfully tested with a wide range of commercial, shareware, and public-domain TCP/IP packages, e.g., FTP PCTCP, Sun PC-NFS, Novell's LAN WorkPlace and PDETHER, WATTCP based IP programs, NCSA & CU-Telnet/ftp, UMN gopher & popmail, Trumpet WINSOCK, and XFS.

ISPA supports a large set of protocols for communication with other vendors ISDN routers or servers. Among these protocols are LAPB, Frame-Relay, PPP, SLIP, and Cisco-HDLC.

ISPA can communicate at least with the following commercial systems: Ascend Routers, Biodata ISDN Router, Cisco Routers, Conet S2M Router, INS/CLS Banzai ISDN Router, netCS ISDN Router, RzK SLIP Bridge, SGI Indy ISDN 1.0, Spyder Routers, SunLink ISDN 1.0, SunLink ISDN 1.0.2.

ISPA was written for use with ISDN BRI PC cards and has been tested with the Teles.S0, one of the cheapest ISDN cards in Germany, as well as with NCP cards, AVM A1 & B1, Creatix S0/16, Diehl Diva, SCOM & S0Tec, Loewe ISCOM C100, MIRO, mbp Solis, NCP P16 & A, and Dr. Neuhaus NICCY 1000 PC.

The current version of ISPA supports two independent active connections at a time. Alternatively, a connection can use both B-channels for load-sharing. ISPA can be loaded more than once, if more than two simultaneous connections to different sites are desired. Load-sharing can be configured as static or dynamic (bandwidth on demand). Dynamic load-sharing can be used concurrently with a second independent connection.

Load-sharing over two channels is implemented using simple round-robin scheduling, because IP doesn't require the original packet sequence. It's completely hardware-independent. It works in the same way as the Cisco implementation of load-sharing over two X.21 interfaces, and I have tested ISPA with a Cisco Router and two Philips TAs. Load-sharing will not double the performance this way, but you can reach around 13 Kbps.

ISDA shareware by Herbert Hanewinkel is available from ftp.biochem.mpg.de in /pc/isdn.

INAR 1.00, the InterNet Access Router

INAR is a fast, easy to configure freeware package that makes a dedicated IP router out of any 80x86 PC running MS-DOS. To communicate with the network interface hardware, it uses packet drivers that comply with FTP Software, Inc.'s packet driver specifications (Rev. 1.09). If you have more than one computer at your site and want to connect your LAN segment to the Internet and/or other LAN/WAN segments, this package is for you!

INAR's most prominent features include the following:

- Up to eight interfaces
- The router component is written in assembler so it is very fast

- Propagates routing information via RIP (including "poisoned reverse")
- Selective default route propagation
- Interfaces and static routes can be marked as hidden (to RIP) and unreachable
- Transient static routes to provide boot-time default routing until RIP takes over
- Supports static routes with variable subnet masks in the same IP net to ensure economic use of precious Internet IP numbers
- Allows source routes to enforce local routing policies
- Source IP address (reverse route) checking to enhance network security
- Proxy ARP (for all known routes)
- Global broadcast forwarding between subnets of the same IP net
- Can send status messages to a UNIX syslog daemon
- BOOTP forwarding
- ISDN and point-to-point interfaces do not need an extra IP address
- Multiple transmission protocols and dial-in/dial-out links on the same ISDN interface
- A single non-cryptic, easy-to-understand configuration file for all software components
- Comes with packet drivers for the most common Ethernet cards as well as with drivers for ISDN and SLIP/CSLIP
- Includes sample config files for the most common cases

INAR 1.00 is available via anonymous FTP from ftp.fu-berlin.de (160.45.10.6) in a self-extracting LHA file named "/pc/msdos/network/inar/inar-100.exe."

Commercial ISDN Software

MS-DOS/MS-Windows

Microsoft

Microsoft TCP/IP for Windows forWorkgroups. Microsoft's TCP/IP for Windows for Workgroups includes the NDIS 2 protocol to support connecting computers running Windows for Workgroups to computers running Windows NT and Windows NT Advanced Server. Microsoft TCP/IP for Windows for Workgroups does not include any TCP/IP utilities; however, support for Windows Sockets is provided, which allows any Windows Sockets-compatible TCP/IP utilities (including terminal emulators and file transfer programs) to be used.

The only supported interface for Microsoft TCP/IP-32 for Windows for Workgroups version 3.11 is Windows Sockets version 1.1. Support for previous versions of the Sockets specification is not provided. In addition, there is no support provided for Raw Sockets (SOCK_RAW), DOS Sockets, or vendor-specific socket implementations.

The Windows Sockets application programming interface (API) was developed to provide a standard application interface to different vendors' protocol implementations. The official name for the dynamic-link library (DLL) used for Windows Sockets 1.1 support is WINSOCK.DLL. Previous DLL versions that were distributed with Microsoft LAN Manager (for example, WIN_SOCK.DLL and WSOCKETS.DLL) are not supported by Microsoft TCP/IP-32. In order for a Windows Sockets application to function with Microsoft TCP/IP-32, the application must support Windows Sockets version 1.1 (WINSOCK.DLL).

Microsoft TCP/IP for Windows forWorkgroups, Versions 3.11 and 3.11A. Microsoft TCP/IP-32 for Windows for Workgroups is an NDIS 3 protocol that includes the following:

- Core TCP/IP protocols, including Transmission Control Protocol (TCP), Internet Protocol (IP), User Datagram Protocol (UDP), Address Resolution Protocol (ARP), and Internet Control Message Protocol (ICMP). This suite of Internet protocols provides a set of

standards for how computers communicate and how networks are interconnected.
- Support for application interfaces, including Windows Sockets for network programming and NetBIOS for establishing logical names and sessions on the network.
- Basic TCP/IP connectivity applications, including ftp and telnet. These utilities allow Windows for Workgroups users to interact with and use resources on non-Microsoft hosts, such as UNIX workstations.
- TCP/IP diagnostic tools, including arp, ipconfig, nbtstat, netstat, ping, route, and tracert. These utilities can be used to detect and resolve TCP/IP networking problems.
- Support for DHCP automatic configuration.
- Industry standard Windows Sockets 1.1 support for third-party and public domain TCP/IP applications such as NCSA Mosaic.

This version of TCP/IP does *not* do the following:

- Include server-side applications for telnet and ftp.
- Include LPR and Gopher.
- Support an MS-DOS-based interface (you can use Windows Sockets instead).
- Support SLIP and PPP to dial in to the Internet.
- Support NFS (although it will likely be provided by third-party vendors).

Microsoft TCP/IP-32 for Windows for Workgroups is available on the Windows NT Server CD and can be downloaded from: ftp://ftp.microsoft.com/peropsys/windows/public/tcpip

NOTE: For more information on the specific bugs fixed in Microsoft-TCP/IP-32 version 3.11a, query in the Microsoft Knowledge Base on the Qxxxxxx number that precedes the following titles:

Q121317: TCP/IP-32 Version 3.11 Does Not Include Terminal Font
Q120051: DNS Reverse Name Resolution Requests Are Incorrect
Q120052: NBT Query Can Hang Computer or Drop Back to MS-DOS

Q122293: LMHOSTS Lookup Can Cause Intermittent System Pauses
Q119575: TCP/IP-32 Winsock Stops FD_READ Notification
Q119918: Winsock: Accept() Sockets Are Unexpectedly Aborted

NetManage Chameleon MS-Windows ISDN TCP/IP Software

Internet Chameleon is a software package for Windows PCs that allows any user to easily navigate the Internet. The Instant Internet application in Internet Chameleon automatically signs up a user for a new Internet user account and connects you to the Internet within 5 minutes. The application suite includes all the tools you need for exploring the vast resources of the Internet; including document browsing (Gopher, WebSurfer), file transfer (FTP client and server), personal communication (Email, NEWTNews), searching (Archie), terminal emulation (Telnet), diagnostics (Ping, NEWT), and user information lookup (Finger, WhoIs). A consistent graphical user interface makes each of these applications easy to use. Internet Chameleon is designed for mobile, home, or remote users who want dial-up access to the Internet through a modem.

Internet Chameleon is a package intended for dial-up use only, using either the SLIP, CSLIP, PPP or ISDN protocols. NetManage's flagship product, Chameleon TCP/IP for Windows, has the additional capability of running TCP/IP over Ethernet, Token Ring, or FDDI LANs—creating your organization's own TCP/IP network. It also includes TN3270 and TN5250 emulation, Visual Script Editor and Visual Script Player applications, a front end to PROFS/Office Vision electronic mail, LPR/LPD support for printer sharing, and a Domain Name Server. ChameleonNFS includes all the functionality of Chameleon, plus a complete implementation of NFS client and server.

Windows Sockets is an industry standard that specifies how network applications communicate with a protocol stack, typically TCP/IP. The current revision level of Windows Sockets is 1.1. The NetManage TCP/IP protocol stack supports Windows Sockets 1.1.

ChameleonNFS 4.5's new features and applications include the following:

- More than 40 applications
- Six integrated networking suites
- FTP firewall support

- VT320 and Wyse emulation
- Host connectivity/terminal emulation—TN3270 and TN5250
- Script recorder
- Autoscaling fonts
- Toolbar
- WinHLLAPI interface
- Telnet
- VT320 support
- SCO ANSI
- Script recorder
- Wyse 50/60 emulation
- Direct serial port access
- FTP and Email interface
- Session Manager—manage multiple sessions from a single application
- Email—draft, template, sent mail folders, spell checker, return receipts on message delivery or opening, folder list integrated in main window, expanded editor, multiple address books
- New applications: Sound Player, Graphics Viewer, Calendar/Scheduler, Electronic Post-It Notes
- NEWTnews—threads, off-line reading
- WebSurfer WWW client
- FTP
- Multiple firewall support
- Macros/scripting with recorder for automated transfers
- NFS
- Support for membership in multiple groups
- UNIX umask support
- NEWTScan—scanner server
- Desktop Management
- Name/Address Resolver
- NIS Lookup

- PC network time—synchronizes PC clock with central network clock
- Rcommands (rcp, rsh)
- UNIX to Windows and Windows to UNIX file conversion utility
- NEWTShooter—instant data exchange between applications
- TCP/IP Protocol Stack
- NEWT—TCP/IP Stack
- Dial-on-demand with automatic disconnect
- Multiple default gateways
- InetD
- DHCP client
- NIS client
- OEMSETUP for ODI, NDIS and NFS
- SNMP support for future automatic upgrades and software distribution
- WinSNMP support
- Automatic installation on any Windows sockets stack
- Interactive log window for manual logon and bypassing of scripting
- Windows interface for building scripts
- 100 modem setup strings

System Requirements
- Software: Windows 3.1, running enhanced mode
- 256-color video driver recommended for use with WebSurfer Hardware
- 386 CPU or later
- RAM: 4MB
- Disk space: 7MB
- 1.44MB disk drive
- Modem (14.4K baud or above recommended)

Combinet EVERYWARE Connection Manager

Connection Manager is a Windows-based application that provides for centralized call processing and distribution, as well as the configuration, management, authentication and accounting for Combinet's on-demand remote access products. With Connection Manager, an enterprise network supports more remote users per enterprise access unit, significantly reducing hardware investment. Connection Manager also provides centralized security and call logging.

Connection Manager benefits include the following:

- On-demand networking, resulting in minimal telephone charges
- Compatible with standards-based ISDN BRI, Switched-56 telephone services, and Ethernet LANs
- Simple installation and maintenance
- Interoperable with all Combinet products
- Computer, network operating system, and application transparency
- Significant enterprise network cost savings
- Centrally administered authentication and callback security
- Call logging

Combinet products include a limited one-year hardware warranty and free software updates for one year.

Features

- Microsoft Windows user interface
- Access units at enterprise are referenced by name

Enterprise Unit Management

- Quick, transparent response time relative to ISDN/Switched-56 call setup time
- IP protocol access unit control messages can operate through routers
- Selectable priorities of allocation pool access units
- On-line status of all access units at the enterprise

Security
- Password required to access Connection Manager console
- Centralized administration of call information
- Password
- Remote unit Ethernet address
- Ringback number

Options for Caller Information
- Call event logging
- Records easily and quickly accessible through database
- Call event record
- Alerts

Specifications
- Enterprise access units supported
- 1000 units in allocation pool
- ISDN BRI: Everyware 200, 400
- Remote access units supported
- ISDN BRI: All
- Switched-56: All
- 10,000 User ID information records

Requirements
- Dedicated 486 33 MHz minimum (486 50 MHz recommended), 8 MB RAM, 100 MB free disk space
- Ethernet card: NDIS compliant driver
- Microsoft Windows 3.1 or later, DOS 5.1 or later

Recommended Hardware
- Remote sites: Any Combinet access unit (ISDN or Switched-56) with Everyware Software Release 2.3
- Enterprise: Combinet Everyware 200 or 400 access unit with Everyware Software Release 2.3 or later.

Windows NT

Microsoft TCP/IP for Microsoft Windows NT version 3.5

Microsoft's TCP/IP for Microsoft Windows NT version 3.5 has been upgraded to include the following features:

- New (completely rewritten) core TCP/IP protocols, including Transmission Control Protocol (TCP), Internet Protocol (IP), User Datagram Protocol (UDP), Address Resolution Protocol (ARP), and Internet Control Message Protocol (ICMP). This suite of Internet protocols provides a set of standards for how computers communicate and how networks are interconnected.
- Support is also provided for Point-to-Point Protocol (PPP) and Serial-Line IP (SLIP), protocols used for dial-up access to TCP/IP networks, including the Internet.
- Support for application interfaces, including Windows Sockets for network programming, remote procedure call (RPC) for communicating between systems, NetBIOS for establishing logical names and sessions on the network, and network dynamic data exchange (Network DDE) for sharing information embedded in documents across the network.
- Basic TCP/IP connectivity applications, including finger, ftp, lpr, rcp, rexec, rsh, telnet, and tftp. These utilities allow Windows NT users to interact with and use resources on non-Microsoft hosts, such as UNIX workstations.
- TCP/IP diagnostic tools, including arp, hostname, ipconfig, lpq, nbtstat, netstat, ping, route, and tracert. These utilities can be used to detect and resolve TCP/IP networking problems.
- Services and related administrative tools, including the FTP Server service for transferring files between remote computers, Windows Internet Name Service (WINS) for dynamically registering and querying computer names on an internetwork, Dynamic Host Configuration Protocol (DHCP) service for automatically configuring TCP/IP on Windows NT computers, and TCP/IP printing for accessing printers connected to a UNIX workstation or connected directly to the network through TCP/IP.
- Simple Network Management Protocol (SNMP) agent. This component allows a Windows NT computer to be administered remotely

using management tools such as Sun Net Manager or HP Open View. SNMP can also be used to monitor and manage DHCP and WINS.
- The client software for simple network protocols, including Character Generator, Daytime, Discard, Echo, and Quote of the Day. These protocols allow a Windows NT computer to respond to requests from other systems that support these protocols.

Macintosh

PlanetPPP for Macintosh

The PlanetPPP software offers high-speed Internet and AppleTalk connectivity over PPP for Macintosh users. PlanetPPP can be installed and configured in less than 5 minutes. The Installer places necessary drivers into the System's Extension folder and creates the PlanetPPP folder on your hard drive. You start the PlanetPPP application by double-clicking on it, configure it for use with your Planet-ISDN board, and enter appropriate security and dialing information in the settings document. Click "Connect" and in a matter of seconds (literally) you are connected with your ISDN-capable Internet Service Provider at 64 Kbps (56 Kbps for those individuals without Clear Channel 64 Kbps access). PlanetPPP seamlessly works with MacTCP. Hence, once the PPP link is established and authorization approved, you are able to use any IP-based application (Mosaic, Eudora, Fetch, Netscape, Telnet, etc.).

Features

- Single B channel synchronous HDLC PPP connections using the Planet-ISDN board
- PAP and CHAP security options
- Multiple settings documents for connections to various hosts
- Both IPCP (Internet) and ATCP (AppleTalk) protocols
- Idle time out
- Connection status window
- Specification of Maximum Transmit Unit (MTU) size
- Link statistics
- Address and protocol compression

PlanetPPP will be shipping with Planet-ISDN boards sold in the USA in the first quarter of 1995. Previously registered users of the Planet-ISDN board can obtain the software for a nominal fee from authorized resellers. Planet is targeting the second quarter of 1995 for the first upgrade of PlanetPPP for the Macintosh. The most important feature of this upgrade is support for Multilink PPP (MLP).

VMS

DEC VAX ISDN Software V1.1

The VAX ISDN software controls the ISDN signaling channel (D), and therefore controls the establishment of the two bearer channel (B) connections over the ISDN to send and receive calls to and from separate destinations. As each channel is independently managed, two different protocols can be simultaneously run on both channels. The DIV32 driver is included in the VAX ISDN software distribution kit. Any DECnet node running ULTRIX, VMS, or MS-DOS has the ability to use the ISDN circuit once the connection has been established.

There are two software versions: VAX ISDN software V1.1, and VAX ISDN Access software (optional). The VAX ISDN software runs on the Q-bus MicroVAX hosting the DEC ISDN controller 100. DECnet VAX and VAX P.S.I. layered network software, as well as customer-developed protocols (HDLC, SDLC, DDCMP oriented), are supported.

Public ISDN networks and switches currently supported are listed in Table 9.1 on page 184.

The following processors are supported:

MicroVAX II 3300/3400/3500/3600/3800/3900

VAXstation II 3200/3500/3620/3540

VAXserver 3300/3400/3500/3600/3602/3800/3900

VAX 4000 Model 300

Distribution media—tape: 9-track/1,600 bpi magtape (PE), TK50 streaming tape

Country	ISDN Network/Switch
United Kingdom	British Telecom ISDN 2
France	Numeris—VN2 Network
Germany*	Deutsche Bundespost—1TR6 Network
Japan	INS-NET-V2 Nippon T&T Corp.
Switzerland	Tested: SwissNet1—1TR6 Access
United States	AT&T 5ESS-5E4 Switches

Version 1.1 enables you to take advantage of the semipermanent mode of the ISDN German network.

Table 9.1 *Supported public ISDN networks and switches.*

Suggested Hardware

The DEC ISDN controller 100 (DIV32) provides Digital systems with Basic Rate Access to the ISDN. It is a single-board, synchronous communication controller, whichcan be fitted directly into a Q-bus MicroVAX system enclosure. Each module supports one ISDN Basic Rate Access line (two high-performance bearer channels and the signaling channel, 2B+D). The board has a 68000 microprocessor and 128 KB of on-board memory to downline load the Level 1 software.

Prerequisite Software

Host: VMS operating system V5.2-1
VMS Tailoring optional software: VAX P.S.I. V4.3, DECnet VAX V5.2
VAX ISDN Software order codes:

Option	Order Code
Software license	QL-VZ9A*-**
Software media	QA-VZ9A*-**
Software documentation	QA-VZ9AA-GZ
Software product services	QT-VZ9A*-**

Table 9.2 *VAX ISDN software order codes.*

Ordering Information

Refer to the following for further information on supported processors and services: Software Product Description 31.23.01

SUMMARY

Although there is a variety of software products available for dealing with ISDN devices and connections, most individual users will only require a TCP/IP stack or packet driver and a PPP communication program. These may be obtained separately or within a package such as NetManage's Chameleon. The most important thing you need to remember about these software packages is that they must be compatible with your ISDN device and your ISP's ISDN hardware and software. First check with your ISP to determine what software you will need to be compatible with theirs. Next check the software provided with your chosen ISDN device to insure its compatibility. If you need additional software, such as a TCP/IP stack, ask your ISP for assistance in choosing the correct package for your system.

If you are using a PC with Windows for Workgroups, the MS TCP/IP-32 stack should work with your device. Both Windows95 and OS/2 Warp contain the appropriate drivers for use with most ISDN devices discussed here.

It's not really as complicated as it appears from the discussions here. Following the step-by-step instructions in later chapters and working closely with your ISP should make installing and configuring your ISDN software as painless as possible.

Getting Started with ISDN

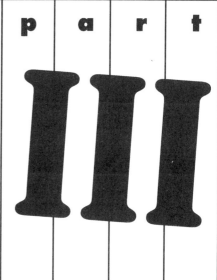

part III

Once you've made your selection of the appropriate ISDN hardware and software for your PC, only then does the real fun begin. Now, it's time to get serious! You'll need to arrange to have ISDN service installed, handle the wiring from the phone company's demarcation point at your home or office, and obtain all the necessary cables and equipment. Only then can you get down and dirty and begin the installation process on your PC or network.

We begin this long, arduous, and sometime painful process with the most basic of all questions: Should you do it yourself, or hire someone else to install ISDN for you? In Chapter 11, we debate the pros and cons of flying solo versus bringing in a real professional. The remainder of Part III is devoted to stepping through the ordering and installation processes involved in bringing ISDN to your PC or your network. Even if you don't do it yourself, you'll find these chapters useful and interesting, because they'll help you understand what your expensive hired gun is up to, and can point you at some things to request, some things to demand flat-out, and others to avoid at all costs!

We continue the process toward working ISDN in Chapter 12, as we review the steps and requirements for a home or small office standalone ISDN installation. Here, we review the elements of ordering an ISDN connection, finding an ISDN-capable service provider, installing and configuring the necessary wiring and termination equipment, and more. In Chapter 13, we cover the same ground from a different perspective, as we lead you through the mechanics of a small-office installation, with the added excitement of integrating ISDN with a LAN.

Chapter 14 proceeds to discuss common ISDN troubleshooting tactics, tips, techniques, and approaches that you might find helpful if and when your ISDN experience hits a snag. Chapter 15 concludes Part III (and the subject matter of the book) with an overview of the most common questions and answers on ISDN subjects, culled from the Frequently Asked Questions lists (FAQs) from several ISDN newsgroups and mailing lists.

Our goal in Part III is to step you through the process of ordering and installing a simple ISDN setup, and then cover the basics for troubleshooting your installation and answering the most common questions you're likely to have. We want you to feel comfortable with the process, and to understand not just the steps involved, but the perils and pitfalls you're likely to encounter along the way. Let us close this introduction to Part III by wishing you a quick and painless installation, or at least, an all-knowing and infallible source for support when you need it!

ISDN Telephones and Business Equipment

chapter 10

The primary focus of this book is to provide information about ISDN, for personal or business use by an individual. The previous chapters dealt with low to moderately priced equipment necessary to connect your computer, any other ISDN S/T device, and in most cases, an analog (POTS) telephone to your ISDN line. This chapter presents a brief look at some of the other possibilities for ISDN, not only for you as an individual, but for your small office of a few people, computers, phones, faxes, and the like.

Although the idea of replacing your home phone system with ISDN phones is appealing at first thought, after you find out that you don't just attach a few $15 phones to a standard series run cable wire pair and have them work, you'll quickly change your mind. Switching your small office over to ISDN may be more appealing, since business phone rates are generally much higher than residential and the higher cost of the ISDN phones isn't out of line with small PBX prices. Office wiring is generally easier to change to ISDN-compatible, if that's necessary. You'll be able to read more about ISDN phones later in the chapter.

Perhaps you are thinking about the more exotic realms of video conferencing and wireless or satellite equipment. You'll even find out about this, and where to go for more on-line information, in this chapter. The appendices contain the most up-to-date sources of information we could find on these topics. Again, we suggest you jump on the Internet and visit the sites listed in the appendices and follow their pointers to the latest and greatest products before you make your final decision to purchase any of them.

ISDN Telephones

An ISDN telephone differs only slightly from the digital phones that most businesses use with their PBX (public branch exchange) systems. However, businesses usually have a communications consultant, or access to someone from the company from which they purchase or lease the equipment, who installs, configures, and maintains their telephone system. An ISDN phone plugged into the S/T jack of your ISDN T/A is going to make your system "completely digital," but is it really going to help you do your business better?

It's certainly going to cost you much more for the ISDN phone ($300 and up) than your standard telephone. Most local telephone service providers are still trying to figure out how to make their switches and software work with ISDN phones to provide the various call features that you will probably expect from any business phone or even your home phone. Count on a lot of wasted time and calls to the phone company (via your analog phone) while getting your ISDN phone up and working the way you expect it to work. In the future, when ISDN phones become the norm, and their price drops accordingly, you may be more interested in checking them out again.

As if this isn't enough to dissuade you, keep in mind that to get an ISDN phone working, you must plug it into an NT1. These digital phones can't be daisy-chained (serially wired) like your current analog phones. You need to run a cable directly from your NT1 or T/A to each ISDN phone. Remember, too, that an ISDN phone is just an S/T device to the NT1 or T/A, and each S/T device is a separate entity. You can't just pick up a couple of handsets on ISDN phones in separate rooms as you can with your analog phones in your home. You have to treat them like the phone system in your office, where you must initiate a conference call to link any two phones together. This isn't because they have different phone numbers; it's because they are handled differently by their digital switching and routing equipment. Chapters 12 and 13 discuss the problems of wiring houses and offices for ISDN usage.

If, after all these caveats and potential gotchas, you're still interested, here are brief descriptions of some of the ISDN phones and phonelike devices currently on the market.

AT&T 8520T ISDN Voice-Data Terminal

AT&T's 8520T voice/data terminal is an advanced speakerphone that enables the user to display a personal directory and other information on a seven-line display screen. The terminal includes 20 buttons for instant access to multiple phone lines or calling features.

In addition, 10 "soft keys" may be programmed to help manage a personal directory, to select among eight distinctive ringing patterns, or to activate other features. Dedicated feature buttons simplify such functions as call transfer and redial, and four control keys make it easy to scroll through the 144-entry personal directory.

When connected to a personal computer, the terminal handles data speeds of up to 64 Kbps and can accommodate simultaneous voice and data communications on a single ISDN telephone line. The 8520T also may be programmed to display information on the computer screen to identify incoming calls, alert the user to priority calls, and keep a log of incoming and outgoing calls. The 8520T voice/data terminal will be available beginning in the second quarter of 1994 with an manufacturer's suggested retail price of $1010.

AT&T National ISDN-2 Terminals

Two new terminals are designed to take advantage of National ISDN-2 (NI-2) technology, which adds new capabilities to the ISDN network. As NI-2 switching capabilities become available in many areas, the new terminals will equip telephone companies and Centrex administrators to download pre-programmed features to each terminal instead of programming individual telephones on-site.

AT&T is offering its NI-2 terminals in 14-button and 35-button models, with such features as full speakerphone, display, menu "soft keys" for call management, and a message waiting light. They will be available in the second quarter of this year. The NI-2 terminal will work in existing NI-1 ISDN installations but will be limited to NI-1 capabilities. Likewise, NI-1 terminals will work in NI-2 installations, but without the added NI-2 features.

AT&T ExpressRoute Digital Adapter 2000

The ExpressRoute Digital Adapter 2000 connects analog and digital equipment used in home offices, small businesses, and college dormitories to an ISDN line. The adapter's two data ports can connect personal computers or workstations at transmission speeds of up to 64 Kbps. Both data ports support D-channel packet services, or a single port can transmit circuit-switched data. An analog port provides access to ISDN voice services and digital conversion for telephones, Group III fax machines, answering machines, and modems.

The adapter is designed to accommodate the telecommuter or home office user without the need to modify existing residence telephone service. Users benefit from its capability to handle high-quality voice and data communication simultaneously. The adapter's advanced voice features can accommodate multiple incoming calls and support ISDN voice features such as conference calling, call forwarding, and Incoming Caller Identification (ICLID).

The adapter is packaged with a network terminating unit, which serves as the interface between the ISDN line and customer premises equipment. Manufacturer's suggested retail price is $850 for the package and $650 for the adapter alone.

PRI

There are three ways ISDN can be "delivered" from an ISDN-ready digital switch. In the first method, ISDN can be delivered through a direct BRI connection from an ISDN switch. Secondly, one or more BRIs can also be linked to ISDN Centrex service. This arrangement offers several advantages for an individual or company. Since the ISDN switch functions as their switching system, the company does not have to own or maintain a PBX or key system. It also offers a low-cost, virtually unlimited growth path. The third method is through a PRI connection.

A PRI delivers 23 B channels plus one D channel from the telephone company to a PBX, ISDN PRI router, or other control device, which then distributes the B channels as needed throughout an organization. The configuration of this setup can vary greatly. Users with heavy data traffic might configure the connection through an ISDN router, multiplexer, or controller, rather than a PBX, thereby reducing the chance of congestion through the switch.

Dynamic allocation of B channels in a PRI is possible now. For practical purposes, though, combining multiple channels in a PRI—for large video conferences, data transfers, and the like—is most often programmed into the digital switch serving the location. However, new bandwidth-on-demand controllers have begun to enable a network manager to combine larger bandwidths in real time to meet specific needs. They can also monitor quality and traffic on both corporate leased-line and ISDN networks, and perform dynamic allocation of B-channels to relieve bottlenecks or backup error-prone or damaged lines.

Installing and configuring a PRI adapter is not something an individual usually attempts in the do-it-yourself mode. A communications consultant with ISDN experience can usually save you money by letting you continue working to make money while the consultant efficiently installs and configures your ISDN system. Most consultants work closely with your local telephone company to insure the smoothest possible transition and least down time for you. If you insist on jumping into it yourself, here are a few PRI devices you should consider.

IBM ISDN Primary Rate Adapter

The ISDN Primary Rate Adapter provides 23 data channels of 64 Kbps, each operating simultaneously, and a 64 Kbps signaling channel for communicating over an ISDN Primary Rate Service. Each channel operates full duplex. By providing digital communication over twenty-three 64 Kbps channels, many individual basic rate ISDN channels can be serviced over a single telephone company connection line. This adapter and associated software will operate with both the AT&T 5ESS and Northern Telecom DMS-100. Price: $7995.

Twenty-four 64 Kbps channels over a T1 communication line are supported. Twenty-three of the channels are B channels for data. The 24th channel is a D channel which provides signaling and control for the B channels. By providing digital communications on twenty-three 64 Kbps channels, many individual basic rate ISDN terminals can be serviced over a single telephone connection line into the business premises. The basic rate terminals gain access via the public switched telephone network.

A Port Connection Manager is also provided as a part of the included adapter software. The included device driver provides an ANDIS- compatible MAC layer of support. These two together allow this product to be used as a wide-area communications adapter by the IBM LAN Distance family of products.

The IBM ISDN Primary Rate Adapter, in conjunction with IBM LAN Distance, further enhances the capability of extending the "office LAN" to remote users. Using switched Basic Rate ISDN, or switched 56 Kbps service, remote users can access the corporate LAN as though they were physically connected to the LAN with performance and response time approaching that of an office terminal connected locally to the "office LAN."

Minimum System Requirements

- IBM PS/2 models 8595 or 9595
- 750kb RAM for LAN Distance Support
- 1.5MB HD space
- VGA Display Monitor
- 3.5 inch 1.44 diskette drive

Software Requirements

- IBM OS/2 2.1 or greater
- The IBM LAN Distance Connection Server Version 1.1 for applications that will provide remote access to the "office LAN"
- A maximum of three ISDN Adapter cards can be physically placed in the PS/2 model 8595 or 9595. The programs included with the IBM ISDN Primary Rate Adapter can support up to three Adapters.
- Warranty period: One year

ISDN Systems, Inc.: FX-PRI ISDN/Frame Relay PC Adapter

The FX-PRI is the first adapter for the Windows NT environment with an integrated smart T1 CSU and the ability to run ISDN-PRI or frame relay drivers. This unrivaled flexibility enables today's Frame Relay device to be upgraded to tomorrow's ISDN-PRI device simply by installing a new driver—with no change to the hardware. Or just add additional FX-PRI adapters, and use your existing hardware platform for both ISDN-PRI and frame relay network connections.

With the FX-PRI adapter, an organization can customize Windows NT-based products for the specific characteristics of their individual connectivity requirements. ISC's FX-PRI is the ideal solution for organizations requiring remote LAN access. Network service and Internet service providers will find that with the FX-PRI they can build very powerful access gateways at a fraction of the cost of devices they currently use.

The FX-PRI can be equipped with both ISDN PRI or frame relay drivers for Windows/NT (NDIS 3.0). By utilizing the NDIS 3.0 driver, the FX-PRI converts your NT RAS into a powerful remote LAN access server for remote ISDN and/or frame relay users. With the FX-PRI, Windows NT may also be used as a powerful application server.

Platform
- MS Windows NT 3.5 and higher
- ISA/EISA bus

Features
- Up to three FX-PRI cards per system
- Three levels of congestion control (supports CIR)
- T-1 and F/TI (4 wire) interfaces
- NDIS 3.0 drivers
- Supports maximum PVCs (1 to 1024)

Hardware
- The PC adapter is designed for the IBM PC-AT bus (ISA and EISA) with:
- On-board Smart T1 CSU (RJ-48)
- ESF or SF
- B8ZS or AMI
- FDL (ANSI or ACCUNET)

Interoperability
- Frame relay—RFC-1490 ensures compatibility with the leading routers
- ISDN—PPP and Multilink PPP ensure compatibility with leading ISPs

Frame Relay Standards
- Frame relay based on ANSI T1S1
- Full support of LMI and Annex "D"
- Protocol encapsulation via RFC-1490

ISDN Standards
- National ISDN-2 and CCITT
- Custom AT&T
- Custom NTI

Protocols
- TCP/IP and IPX Routing
- NetBIOS Gateway
- 802.3 and 802.5

Primary Rate Incorporated PRI-ISA48 Dual T1/ISDN Controller

If your application requires access to any T1-based service, Primary Rate Incorporated has both the hardware and software that you need. The PRI-IAS48 Dual T1/ISDN Controller is an Industry Standard Architecture (ISA)-based T1 subsystem for data and/or digital voice communications. It contains an ISA slave interface for T1 to host computer communications with HDLC and DMA hardware to format all combinations of channelized and nonchannelized data for up to 32 full-duplex data streams. It also includes a standard Multi-Vendor Integration Protocol (MVIP) bus interface that provides a multiplexed digital telephony highway for adapter board communication within a personal computer chassis. The board comes with two T1 interfaces, and complete on-board signaling and data formatting software.

The PRI-ISA48 allows your system direct access to the power and speed of T1, fractional T1, and primary rate ISDN lines. Its flexible architecture makes it also ideally suited for use in both frame and cell relay environments. The PRI-ISA48 board comes complete with the most extensive software available for any board of its kind.

The Q.931 ISDN software module provides the means to establish, maintain and terminate network connections across an ISDN between communicating application entities. It provides generic procedures which may be used for the invocation and operation of supplementary services. PRI s Instant ISDN Software Q.931 is capable of controlling both circuit-switched and packet-switched connections. Q.931 provides its services to the upper layers via PRI's Simple Message Interface (SMI). This interface consists of two wrap-around queues. One queue (L4L3) provides service requests from the higher layers to Q.931. The second queue (L3L4) provides service indications from the Q.931 to the higher layer. To perform its functions, Q.931 utilizes underlying resources such as PRI's Instant ISDN Software Q.921 to provide network connections.

Instant ISDN Software (LAP-D)

The Q.921LAP-D software conveys information between the Q.931 entities across the ISDN user–network interface using the D-channel. PRI's LAP-D includes support for multiple terminal installations at the user-network interface and multiple layer 3 entities. All data link layer messages are transmitted in frames which are delimited by flags. (A flag is a unique bit pattern.) Each frame has a CRC-16, prior to the closing flag, which may be used to detect the occurrence of one or more corrupted data bits. Both flags and CRC-16 are typically generated and interpreted by a hardware device known as an HDLC controller.

VIDEO CONFERENCING

Early video conferencing utilized large pieces of expensive equipment to provide "room"- based video conferencing. The paradigm for room-based video conferencing is that participants at a site all gather together in a specially equipped conference room around a conference table and look at monitors displaying similar rooms at remote sites. Desktop video conferencing is a new paradigm for video conferencing. It is "desktop"-based with participants sitting at their own desks, in their own offices, and calling other participants using their personal computer in a manner much like a telephone.

Bandwidth is the limiting factor associated with communication. The limiting factor in a system is referred to as the bottleneck. Sending video through a communications channel requires a lot of bandwidth. In theory, since video conferencing supplies a continuous stream of data into a channel and packet switched communication does not guarantee that the bandwidth required is available, circuit-switched communication is the better of the two for video conferencing. There are drawbacks, though. ISDN is currently the only widely available form of circuit-switched digital service. A standard ISDN connection, which is affordable for the desktop, can provide 128 kilobits per second of bandwidth. Full motion video is 30 frames/sec. ISDN can achieve frame rates of about .25 frames/sec., or one new frame every 4 seconds. With some data compression algorithms and very fast Pentium PCs or Macintoshes, full-motion video conferencing that produces acceptable, but jerky, black-and-white video, is possible over ISDN lines.

General Description: ISDN Video-conferencing Products

IBM

IBM's Person 2 Person is a PC-based product which offers desktop video conferencing using Ethernet, ISDN, or serial connections. It requires Microsoft Windows for Workgroups to utilize ISDN, or just Windows for the other connection schemes. It also provides a shared whiteboard facility. The hardware requirements are steep. Person 2 Person requires an Action Media II with Capture Option for video capture and playback, and an IS-COM ISDN board for ISDN connectivity.

Intel

Intel's ProShare is a software/hardware bundle that provides desktop video conferencing over ISDN and Ethernet. Experience with ProShare so far has been extremely positive. It provides a shared whiteboard facility based on a notebook metaphor, and it has a well-structured interface. The product support for ProShare has been very positive. ProShare can also operate (without video) over a serial link. There is a scaled-down version of the product that provides all the functionality of the full version but with no video/ISDN board. There is also some movement by Intel to standardize the video format, which may eventually provide interoperability across platforms and software products.

Sun

Sun's integrated package, ShowMe, provides a complete video, audio, shared whiteboard, and shared application system that runs on Solaris 2.3 UNIX systems. It is extremely configurable and has a clean, well-designed interface. It runs with an IP protocol so that it can operate across the Internet. Reception of video can be done without any special hardware, just the ShowMe executable. Video capture requires a video board and camera.

MBONE

The MBONE is a virtual network implemented as a subset of the Internet. It uses the IP-multicast protocols to provide multicast video, audio, and shared whiteboard facilities across the Internet. MBONE provides multipoint connections, either one-to-many or few-to-few, while preserving Internet bandwidth by making use of multicasting. The MBONE programs (nv, vat, wb and sb) run on UNIX workstations. They are freely available on the net.

CU-SeeMe

CU-SeeMe is an experiment in providing real-time delivery or video and audio signals across the TCP/IP Internet. It was created at Cornell University. It provides one-to-many connections using UNIX reflector software. It does not use multicast, so the bandwidth it consumes increases with each connection to the reflector. There are plans to make it compatible with MBONE. The software is freely available across the net and from Cornell's Web site, http://www.cornell.edu/.

ISDN VIA WIRELESS AND SATELLITE EQUIPMENT

Several consultants are available in the ISDN wireless and satellite transmission field. They generally offer assistance in wireless extensions for ISDN 2B+D on VSAT Ku-Band space segment, ISDN 2B+D on RF, and broadband ISDN on submarine fiber optic cable, point-to-point and point-to-multipoint applications for voice, video, data and TCP/IP. They offer assistance in broadband ISDN between North America and Europe on the new submarine fiber optic super cable CANTAT-3 as well as wireless Internet Points-of-Presence (POP) servers.

A few of the products starting to show up on the market in this area are briefly described below. Check on Dan Kegel's WWW site for the latest information.

ISDN/Universal Wireless Converter D.I.C.A. 6400

TeleMark's D.I.C.A. 6400 ISDM/Universal Wireless Converter provides low-cost wireless extension of ISDN interfaces. It is fully compatible with the D.I.C.A. 6100 and D.I.C.A. 6200.

Features
- Multiple BRI Transmission
- Up to 14 wireless ISDN extensions
- Full support of all ISDN types worldwide
- Full support of any wireless modems

ISDN Multiplexing
- Transmission of up to four Basic Rate Interfaces over one wireless pipe
- Various options for configurations

Wireless ISDN-on-demand via Satellite
- On demand via reservation
- Dial-up on demand service

Remote Control
- Remote configurable
- Remote status report

Interfaces
- To wireless modem via RS-449 or V.35
- To ISDN network and applications via ISDN Basic Rate S/T Interface
- To remote control RS 449

ISDN/Satellite Converter D.I.C.A. 6100

TeleMark's D.I.C.A. 6100 ISDN/Satellite Converter (ISDNSat) provides low-cost satellite extension of ISDN Basic Rate Interfaces. It upgrades any existing satellite earth station with ISDN capabilities 2 B + D, 1 B + D.

ISDNSat provides the following:
- Unlimited distances for voice, data, and fax transmission
- Internetworking, video conferencing via ISDN
- Full support of ISDN signaling and functionality
- Full support of all types of ISDN worldwide
- Full support of all types of satellite modems

Requirements
- Your existing satellite equipment
- Your existing ISDN equipment
- A satellite space segment with 160 Kbps data rate for 2B+D or 64 Kbps for B+D

Interfaces
- To satellite modem via RS-449 or V.35
- To ISDN network and applications via ISDN Basic Rate S/T interface

ISDN TEST EQUIPMENT

Unless you are going into business as an ISDN installer or systems consultant, you will probably not need any serious ISDN test equipment. Most BRI devices (adapter cards, modems, etc.) are supplied with testing software for your PC. This software tests the ISDN device and ISDN line all the way to the phone company's switch. If it finds something wrong, you generally call the ISDN service provider (phone company) and report the problem. Then let them use their expensive test equipment to find the cause of the problem. The following brief description and price of one BRI test device should let you see why you probably won't want one.

The UPA 100 BRI Protocol Analyzer

Until now, ISDN protocol analyzers have been prohibitively priced from $8000 to over $35000. This has made it difficult to justify the purchase, especially if you just need it occasionally. However, without a protocol analyzer it is almost impossible to resolve the "finger-pointing" that commonly occurs between the service provider and the equipment vendor.

The UPA 100 is a compact unit (9.5" x 7.5" x 1.5") that connects to the serial port of an IBM-PC compatible or Mac computer. It captures the messages passing between the subscriber's equipment and the network, decodes them into plain English, and displays them on the screen and places them in a buffer for later analysis. Circuit-switched and packet-switched decodes of National ISDN, Northern Telecom DMS 100 Custom, and AT&T #5ESS ISDN are supported. At $2495, the UPA 100 may be the lowest cost Basic Rate analyzer on the market.

Summary

As you can see by this brief overview of ISDN equipment, you'll probably be well served in your home or one-person office by your PC with an ISDN T/A and NT1 with an analog phone plugged into it. If you really want to use an ISDN phone with an external NT1 and UPS, there are several good, but expensive, ISDN phones available.

If you want to convert a small office with several people to ISDN, your best bet will be to hire a communications consultant to help you with it. You can probably make more money doing whatever you normally do and paying a consultant than you could save by wasting your time trying to do it yourself.

chapter 11

MAKING GOOD ISDN CHOICES

When it comes to dealing with ISDN—especially configuration and installation issues—there's a fundamental decision you'll need to make: Are you going to do it yourself, or enlist the aid of a real ISDN professional? Most of the contents of this book has been created to help you take either route, by understanding requirements, costs, and equipment well enough to work intelligently with ISDN. In this chapter, we'll tackle your methods for ISDN deployment head-on, and provide some input about how to find and work with a consultant, or how to go it alone with a minimum of stress and strain.

To make the correct choices regarding ISDN, you'll need a bit of a refresher on ISDN terms. Just in case you're like the rest of us and can't remember the acronyms, here's a short list of the most important ones you should know (you can also consult the Glossary at the end of this book for any other terms you might not recognize):

The "Short List" of ISDN Acronyms

- BONDing—B channel BONDing uses both B channels for data to obtain 128 Kbps transmission rates in some TAs.
- BRI—Basic Rate Interface comprising two B channels (bearer channels at 64 Kbps) and one D channel (delta channel at 16 Kbps for telephony signaling information and X.25 packet data).
- ISP—Internet Service Provider.
- POTS—Plain Old Telephone System. Your current analog telephone system.
- PRI—Primary Rate Interface comprising 23 B channels and one D channel with the same physical interface as a T1 circuit.
- NT1—Network Termination device connected to the ISDN line's U interface and providing one or more S/T jacks for ISDN TAs and sometimes one or more POTS jacks.
- RBOC—Regional Bell Operating Company (e.g., Southwestern Bell, Pacific Bell, etc.).
- SLC—Subscriber-Loop Carrier (computerized substation of phone company for ISDN outside the 3.4 mile range of the central office switch; used instead of a repeater).
- SPID—Service Profile Identifier number(s).
- TA—Terminal Adapter device between the NT1 and your ISDN devices or computer.

Do It Yourself or Hire a Consultant?

To decide whether you want to do it yourself or hire an ISDN consultant, you need only answer these three questions:

1. Can you afford to spend a considerable amount of your time talking with your local telephone service provider, selecting and ordering your ISDN hardware, fiddling with your telephone wiring, installing ISDN computer hardware (NT1 and TA), finding an ISDN Internet Service Provider, installing and reconfiguring your computer's Internet software, and troubleshooting your ISDN system over the

3–6-week time period between ordering your ISDN installation and its completion and testing?

2. Are you comfortable installing hardware into your computer, setting it up, wiring your telephone lines, and installing ISDN drivers and Internet software, all by yourself?

3. Do you have more time to spend setting up your ISDN system than you have money to spend hiring a consultant?

If your answer to all of these is Yes, then go for it yourself. Otherwise, find a consultant (or perhaps, two or three of them) and get a bid for a turn-key system tailored to your particular circumstances.

Hiring an ISDN Consultant

For many businesses, hiring an ISDN consultant will be the most cost-effective way to acquire ISDN service. This is especially true if you have more than a couple of people and computers in your business and want to take full advantage of ISDN's capabilities.

Before you start calling consultants from the Yellow Pages, you should answer as many of the questions in the next section as you can about your own company, and about your desired uses for ISDN. Any consultant worth his or her freight will ask you these questions anyway, so you can save time (and money) by having these answers ready. You will be able to better qualify consultants if you know as much as possible about your own desires and needs prior to interviewing them, anyway.

One very good method to use when searching for professional help is to prepare a written request for proposal (RFP). It should include a list of your needs, as well as the questions you want the consultant to answer prior to hiring him or her. You can use the checklist that follows in the next section to help you prepare your RFP.

Keep in mind that since ISDN is a relatively new technology, you will probably encounter a variety of ISDN consultants. Some will have come from UNIX programming backgrounds, some from telecommunications, some straight out of college, and some from who knows where. Some will try to sell you on a complete ISDN telephony and/or computerized system, whether you need it or not. Some will want you to hire them to produce custom programming for your location, whether you need it or not. Ignore these time- and money-wasting proposals.

Fortunately, some consultants will first propose to study your needs, to provide you with hardware and software alternatives that suit your needs and budget, and only then move on to the next steps. After they're sure you know what you need, and how much it costs, they'll proceed to install the chosen system, and to help you and your employees learn to use that system. Hire one of these consultants, but only after you check their references thoroughly and get all specifications and costs in writing. Remember, the job you save may be your own!

CHECKLISTS OF USES, HARDWARE, CURRENT PHONE SYSTEM, ETC.

If you've made it this far into this book, we hope you've already answered the questions from Chapter 5 and have looked over the information contained in Chapters 6 through 10 that are pertinent to your situation. Having decided that ISDN may be what you want or need, you are now ready for the final test. Test? Who said anything about a TEST?? Don't worry, it's open-book with no time limits, and we'll even help you find the answers! In fact, it's more a test of your resolve and your ability to do business, so it's a test you'll definitely want to take!

Ask Yourself:

Which of the following cover your planned uses for your ISDN service?

____ Home
____ Office (non-residence)
____ Personal
____ Business
____ Single user
____ Multiple users
____ Replace existing analog service
____ Digital computer connections and additional voice line
____ ISDN Internet connection for faster access.
____ Use current analog phone(s) with ISDN line

- Use ISDN phone(s)
- Is the telephone wiring at your location a typical analog system with a parallel circuit of the same two wires throughout the building for a single phone number?
- Does your location have an unused pair of telephone wires suitable for ISDN signals for each ISDN BRI you think you will need?

Ask the Phone Company:

- Is your location in an ISDN service area?
- What is the ISDN installation charge from your ISDN telephone service provider?
- What are the monthly charges (basic and usage) for ISDN service?
- Will your ISDN service be via a repeater?
- Will your ISDN service be via an SLC?
- Which central office ISDN switch does your service use? (Siemens, AT&T, etc.)
- Which protocol does the switch use? (i.e., NI-1 [National ISDN-1], AT&T 5ESS Custom, AT&T G3 PBX, or Northern Telecom DMS-100 Custom, Northern Telecom BCS-34 [PVC-1], or other)
- What calling features are available from your ISDN provider? (e.g., EKTS, CACH)
- Does your ISDN provider offer BRI (2B+D) with two SPIDs?
- Does your ISDN provider offer voice service on both B channels on the same BRI service with two SPIDs?

Ask the ISDN Software and/or Hardware Vendor:

- Does your chosen NT1 device work well with your chosen TA and with the phone company's switch and protocol?
- Does your chosen NT1 device have 0, 1, or 2 POTS jacks?
- Does your chosen NT1 device have an internal power supply?
- Does your chosen NT1 device have a battery backup?

____ Does the ISDN TA you are looking to purchase communicate well with the phone company's switch and protocol?
____ Does your chosen TA support B channel BONDing?
____ Does your chosen TA's software support PPP or the protocol your Internet Service Provider offers?
____ Does your chosen TA have multiple S/T jacks?
____ Does your chosen TA have 0, 1, or 2 POTS jacks?

Ask the Internet Service Provider:

____ What NT1 does your ISP suggest?
____ What TA does your ISP suggest to work best with their system?
____ What protocol does the ISP support? (Async. PPP, Sync. PPP, etc.)
____ Does the ISP have experience with connecting your chosen TA to their system?
____ Does your ISP's software support 2 B channel BONDing?
____ What are the extra charges by your ISP for 2 B channel BONDing?
____ What are the ISDN setup charges from your ISP?
____ What are your ISP's monthly ISDN account charges (basic and usage)?

DOING ISDN YOURSELF

In most cases, four vendors will be involved in helping you get your ISDN service completely installed and running: your local phone company, an Internet service provider (ISP), an ISDN software provider, and an ISDN hardware provider. This assumes you already have a 386 (or better) PC running Windows 3.1, WFW 3.11, or OS/2. The rest of this chapter provides a brief overview of the steps you will need to follow to get ISDN up and running.

Step 1. The ISDN Phone Service

You will need to contact your ISDN dialtone provider (the local telephone company; usually RBOCs and independents) to determine if ISDN service is available in your area, and to determine the available options and prices for the service. The following list of ISDN contacts in the USA should help you start your search for the perfect ISDN provider.

National ISDN HotLine   1-800-992-ISDN
Fax: 201.829.2263
E-Mail   isdn@cc.bellcore.com
URL: http://info.bellcore.com
system prompt:ftp   info.bellcore.com

Company	Contact	Telephone No.
AMERITECH	National ISDN Hotline	1-800-TEAMDATA 1-800-832-6328
BELL ATLANTIC In N.J., call your local telephone office.	ISDN Sales & Tech Ctr	1-800-570-ISDN 1-800-570-4736
	For Small Businesses	1-800-843-2255
BELLSOUTH	ISDN HotLine	1-800-428-ISDN 1-800-428-4736
CINCINNATI BELL	ISDN Service Center	1-513-566-DATA 1-513-566-3282
NEVADA BELL	Small business	1-702-333-4811
	Large business	1-702-688-7100
NYNEX	ISDN Sales Hotline	1-800-438-4736
	New England States	1-617-743-2466
PACIFIC BELL	ISDN Service Center	1-800-4PB-ISDN 1-800-472-4736
	24 Hr. Automated Avail. Hotline	1-800-995-0346
ROCHESTER	ISDN Information	1-716-777-1234

Company	Contact	Telephone No.
SNET	Donovan Dillon	1-203-553-2369
STENTOR (Canada)	ISDN "Facts by Fax"	1-800-578-ISDN
	Steve Finlay	1-604-654-7504
	Glen Duxbury	1-403-945-8130
SOUTHWESTERN BELL	Austin, TX	1-800-SWB-ISDN
	Dallas, TX	1-214-268-1403
	North Houston, TX	1-713-537-3930
	South Houston, TX	1-713-567-4300
	San Antonio, TX	1-210-351-8050
	ISDN availability other locations	1-800-992-ISDN
U S WEST	Ron Miller	1-303-965-7153
	Ron Woldeit	1-206-447-4029
	Denver, CO	1-800-246-5226

National ISDN Long Distance Carriers:

Company	Contact	Telephone No.
AT&T	AT&T Front End Center	1-800-222-7956
GTE	Nationwide availability/pricing	1-800-888-8799
	Ron Sterreneberg	1-214-718-5608
MCI	Tony Hylton	1-214-701-6745
	ISDN Availability	1-800-MCI-ISDN
US SPRINT	Rick Simonson	1-913-624-4162
WILTEL	Justin Remington	1-918-588-5069

ISDN Line Configurations

With ISDN comes the ability to select from virtually thousands of configurations for setting up your phone line. Your ISDN service provider will undoubtedly give you a list of the possibilities when you contact them. They would like to sell you as many as they possibly can. Unless you're planning on using an ISDN telephone (EKTS), you don't really need any of the "fancy" ISDN telephony features just to plug your current analog phone into the POTS jack on your NT1 or TA. However, you will need to be sure to get those basic features that you do need.

Depending on which regional phone company serves you, you will probably find that your choice of line configurations is an issue of economics. We recommend that you get only those features you believe you will need. You can always add more features later if you need them.

Typical ISDN Line Features:

# of Channels	Channel Type	Typical Use
(1B)	CSD	64K Internet access (data only)
(1B)	CSVD	64K Internet OR voice
(2B)	CSD+CSV	64K Internet AND voice
(2B)	CSD+CSD	64K or 128K Internet (data only)
(2B)	CSD+CSVD	64K Internet and voice, OR 128K Internet (good choice for 128K Internet and sporadic POTS)
(2B)	CSVD+CSVD	64K data and Voice, OR 128K data, OR two voice lines (best choice if your ISDN provider allows two voice lines)

Definitions:

1B	One B-channel
2B	Two B-channels
CSD	Circuit-switched data on the B-channel
CSV	Circuit-switched voice on the B-channel
CSVD	Alternate voice OR data, on demand on the B-channel
EKTS	Electronic Key Telephone System. A phone with intelligent keys and dialing features normally found on a business phone. Most ISDN phones were designed for businesses and take advantage of EKTS features.
POTS	Plain Old Telephone System. There are special NT1s that will support your analog equipment—such as fax, modem, and analog phone. POTS in this table refers to use of those NT1s.

For individual personal/business use at a residence with a connection to the Internet and a POTS phone plugged into the NT1 or TA, the (2B) CSVD+CSVD 64 Kbps data per B channel line configuration, with only one voice call at a time allowed, should provide good service at a reasonable price.

Step 2. ISDN Internet Service Providers

ISPs across the USA are scrambling to provide ISDN dial-up service. Some are caught between their ISDN line providers (the phone company) and their hardware and software providers (router hardware and accounting software to automatically track connect time). Many ISPs can't get the lines they need from the phone company at a price they can afford. Most can't get the ISDN hardware and software at prices that they can remarket profitably, so that you, the user, won't pay high ISDN Internet access fees to your ISP on top of high ISDN fees charged by the phone company. There's not much you can do about the phone company's fees, but ISP's costs are coming down somewhat due to competition in the ISDN hardware and software market. So shop around before signing up for an ISDN account with an ISP.

If the catchword is to "Shop around," how should you proceed to follow our advice? It's really fairly easy via the Internet. The following URLs can get you to the home pages for the majority of ISPs that offer ISDN connections. You can also ask for help finding the best ISDN ISP in your area on the comp.dcom.isdn USENET newsgroup on the Internet.

Dan Kegel's ISDN Internet Providers WWW site:
http://alumni.caltech.edu/~dank/isdn/

ICUS mirror of the Dan Kegel's ISDN pages:
http://www.icus.com/isdn_ip.html

Internet Service Providers Organized by Services—by CyberBiz Productions:
http://www.cybertoday.com/cybertoday/ISPs/Products.html#ISDN

Make sure that you find out where the ISP is physically located. If the provider is in another town, then you may have to pay long distance charges in addition to their standard connect fees each time you use their service. If that's the case, our advice is: "Keep shopping!"

Step 3. ISDN Internet Access Software

Usually you will need two software resources, a TCP/IP program that connects your computer's TA to the Internet, and one or more application programs for navigating the Internet (FTP, WWW, Telnet, Archie, Gopher, News, Mail, etc.). If you're currently using a set of programs to connect via SLIP to the Internet, you may be able to use most of them in the same manner. However, if you're using a WinSock-compliant program such as Trumpet Winsock on a PC with Windows for Workgroups 3.11 (WFW), you'll probably be better off switching to MicroSoft's TCP/IP-32 software, and using PPP instead of SLIP in your WWW browser, FTP program, etc. This has been discussed more fully in other chapters of this book.

Ask your ISP for help in this area. Many ISPs will provide you with the TCP/IP software, since the majority of it is either freeware or shareware. If you are using OS/2 Warp, your TCP/IP software is built into the operating system and you can use IBM's WebExplorer software directly with it. You can find out more about this package at: ftp://ftp.ibm.net/pub/WebExplorer/web0331.zip.

If you're using a PC with WFW 3.11, you can choose from the following ISDN-compatible software packages, which are all Winsock 1.1 compatible and will therefore work with most ISDN PPP applications as well as the MS TCP/IP-32 software:

Company	Web site (URL)
AIR Mosaic	http://www.spry.com/sp_prod/airmos/airmos.html
Cello	http://www.law.cornell.edu/cello/cellotop.html
InternetWorks	http://www.booklink.com/
NCSA Mosaic	http://www.ncsa.uiuc.edu/SDG/SDGIntro.html
Netscape	http://www.netscape.com/
WinTapestry	http://www.frontiertech.com/
WinWeb	http://galaxy.einet.net/EINet/clients.html
WebSurfer	http://www.netmanage.com/netmanage/apps/websurfer.html

Any of these packages will do the job of letting you navigate the Internet. Specifically, they are WWW browsers, but several contain additional applications such as E-mail, Gopher, and FTP clients. Most have a shareware or freeware test version for you to try. Try as many as you can stand, and use the one you like the best! (For whatever it's worth, we're most partial to Netscape today; who knows what will happen tomorrow?)

Step 4. ISDN Hardware

As we hope you've learned in the previous chapters, you will need an NT1 and a TA for your ISDN system. Ask your ISP and your trusty hardware vendor about the various models they have used successfully. Refer to Chapters 6, 7, and 8 for discussions of ISDN NT1s and TAs, and the methods for determining which types are best for your circumstances.

SUMMARY

So much for the paperwork and preliminary testing. If you filled out the blanks next to the questions at the start of this chapter, you've passed the test. If you have decided to install your own ISDN system and have chosen, at least on paper, an ISDN phone service provider, ISDN ISP, an NT1, a TA, and a suite of Internet access and application software, you're ready to begin the process. Chapters 12 and 13 relate the gory details of a home office ISDN installation and an individual business ISDN installation, respectively. So read on, Macduff!

Getting Your Own Home ISDN System

The time has arrived to show you a typical home installation from start to finish. Having followed the suggestions in the previous chapters, we'll assume you have completed the checklists as follows:

You want to install your ISDN system yourself for your single computer with a single analog (POTS) phone plugged into your system for making calls only when your computer is running. You want the least expensive ISDN hardware for use primarily with your computer to connect to the Internet, and other networks, via your ISDN line. You want full 128 Kbps throughput, but you want the analog phone to ring through and to have one B channel answer it even if you're using both channels for data.

You have found out that you are inside of an ISDN service area (your neighbor has it) and you know from him that the installation and monthly charges are within your budget. You currently only have a single telephone number at your home.

You know from looking at their home page that your ISP offers dial-up ISDN using PPP for a price you can live with. They have no suggestions on hardware but your neighbor is using an ISC SecureLink II card with his 486-66 running WFW 3.11, and if he can do it, you're sure you can do it too. It's now time for you to get into the game and call the first play

Ordering ISDN Service from Your Local ISDN Dialtone Provider

Call your local telephone company's information line and get connected to their ISDN order line, if they have one. Most likely it will be in the "business" rather than "residential" department, but don't let that bother you. Tell the nice person that you want to order an "ISDN BRI 2B+D" line installed at your house ... then wait for the inevitable pause from the phone company person. If he or she doesn't pause, but says, "Certainly, sir, may I have your present phone number?" you've landed a winner who knows what's going on.

After you two determine that you do indeed live in an ISDN service area, get answers for these questions (for the purpose of this discussion, we'll assume you got the following answers):

$150	What is the ISDN installation charge?
$70 flat	What are the monthly charges (basic and usage) for ISDN service?
no	Will your ISDN service be via a repeater?
yes	Will your ISDN service be via a SLC?
Siemens	Which central office ISDN switch do they have? (Siemens, AT&T, etc.)
NI-1	Which protocol does the switch use? (i.e., NI-1 [National ISDN-1], AT&T 5ESS Custom, AT&T G3 PBX, or Northern Telecom DMS-100 Custom, Northern Telecom BCS-34 [PVC-1], or other)
NA	What calling features are available? (e.g., EKTS, CACH)
yes	Does your ISDN provider offer BRI (2B+D) with two SPIDs?
yes	Does your ISDN provider offer voice service on both B channels on the same BRI service with two SPIDs?
same price	What are the prices of 2B - data vs. 2B - 1data/1voice vs. 2b - both voice or data?

Write these answers down so you can refer to them later. If you like what you hear, ask when the service could be installed. There will be a wait while the person checks the installation database and gets back to you. No matter what you are told, you can't do much about it so you either order the installation or you don't.

When you place your order, have the phone company representative confirm the installation price and monthly service for the features you have ordered. You will probably be given an order confirmation number and a phone number to dial to check on the progress of your order. Write these down also. You will probably need them. (We're not pessimistic, just realistic.)

While TPC (the phone company) is getting ready to install your line, or turn it on, you can get ready for the big day by purchasing your equipment, software, and signing up for an ISDN account at your ISP.

Purchasing Your ISDN TA/NT1 Interface Card

If your local hardware vendor doesn't stock the ISC SecureLink II card, have them order one for you. Be aware that the delivery time has been about three weeks. You may be able to get one from a mail-order company such as the ISDN Warehouse or another one you may find via the Internet or in your local area. Ask around, but remember it will probably take TPC upwards of three weeks to install your line anyway. If you have to wait for delivery, go on to the next step below. If you're lucky and find the card locally, buy it, take it home, take it out of the box, look at it, put it back in the box, then give your ISP a ring to make sure you can get that ISDN dial-up account using PPP they promised you last month.

Ordering Your ISDN Internet Service

You did your homework and they are supposed to have the service available at a price that seemed OK at the time. Check the price again and be sure to ask about setup charges and monthly charges and/or connect time charges. Also ask if they offer 2B channel BONDing and what it costs. Write everything down immediately. If everything still looks good, place your order and see when they can have their end ready.

Bear in mind that few ISPs have automated the ISDN setup on their end as they have for SLIP accounts. It may take as much as 30 minutes for their network person to configure an account for you, when they get the magical "round-tuit."

They will probably want to know what TA and NT1 you are using in addition to what software you will use to dialup and connect your PPP link. You tell them you're using an ISC SecureLink II card with built-in NT1 on your 486-66 with WFW 3.11 and the MS Wolverine TCP/IP-32 stack, and they should be happy.

If, for some reason they say their Internet Server (hardware or software) doesn't work well with your planned system, ask if they have a better suggestion. Then listen, write it down, and ask if they will help you install and configure it for free. So far, the system specified above has not been griped about on comp.dcom.isdn and several people have said it works fine for them. Therefore, if the ISP has trouble with it, you may want to call another ISP. If the ISP will help you install and configure another card or software for the same price as the SecureLink II with the same features, you might try it, as long as they guarantee it will meet your specifications and refund your money if it doesn't.

They will probably tell you that they can be ready for you in a couple of days, which is sooner than you need them to be. Tell them the installation date for your ISDN service, and ask if you can call them back when it gets installed and you get your card installed in your computer. They will say, "Sure."

ISDN DIALTONE SERVICE—WIRING, INSTALLATION, CONFIGURATION, AND TESTING

Wiring Your Home for ISDN

In the scenario for this installation, the house has only one POTS telephone currently installed. This leaves a second pair of wires in the normal four-wire cable to use for your ISDN service. Now everything you read will tell you that you should only use twisted-pair wire for ISDN. While this is true, many if not most standard house wires will support ISDN and POTS with no crossover at all. It doesn't hurt to give it a try. Of course you can always buy a few meters of four-wire twisted-pair phone cable at your local Radio Shack or hardware store and run it from your demarc (demarcation point:

the telephone box on the outside of your house where the lines come in) to the room where your computer is located.

You may be a little worried about getting shocked when dealing with your telephone wiring. It's possible, but only when the telephone rings. The voltage is approximately 110 volts, but the current is low, so unless you're wearing a pacemaker, it should just tingle a lot and scare you. So don't do the wiring standing on a ladder unless you don't touch the bare metal of the wires or the screw posts with your bare skin or unshielded metal tools. Use common sense, and you should be just fine.

Whether you use existing wiring or new cable, you'll need to have two wires at the demarc to connect to the ISDN line the phone company installs. You need to put either an RJ-11 or an RJ-45 jack on the other end of the cable next to your computer. If you are using your existing house phone wiring, you can purchase a double RJ-11 wall jack plate to replace your existing single jack plate. Simply wire the existing two POTS wires (usually red and green) to one of the jacks and the other two wires (usually black and yellow) to the other jack, making sure you use the inside (middle) two terminals on the jack. Test the POTS phone to make sure it "breaks" the dialtone and will call out on your existing POTS line.

If you're using a new cable, you can purchase an RJ-11 jack that attaches to the wall or one that looks like the female equivalent of an RJ-11 plug that fits on the end of the cable itself. Most good electronics stores carry at least one of these, if not both.

To test the integrity of the second pair of wires (which you are going to use for your ISDN line), plug your POTS phone into the jack with the second pair. Go out to the demarc and take your current POTS wires (red and green?) off of their connectors and replace them with the other pair (black and yellow?). Go back inside and try dialing out on the phone. If it doesn't break the dialtone, go back outside and reverse the black and yellow wires and try again. If this still doesn't work, you may have a short in the black and yellow wire pair.

You can test this directly with a volt/ohmmeter by twisting the black and yellow wires together at the demarc, after removing them from the screw posts, and using the ohmmeter on the other ends (at the RJ-11 jack inside) to see if you get any resistance. It's really quite straightforward.

After your wiring passes these tests, you're ready for the telephone company to connect the ISDN line. For a comprehensive look at ISDN wiring, contact the North American ISDN Users' Forum (NIUF) at 301-975-2937 or e-mail dawn@isdn.ncsl.nist.gov and request the "ISDN Wiring Guide for Residential and Small Businesses." It will tell you much more than you probably care to know, but who knows, it may make for interesting conversation at your next ISDN users' group meeting.

ISDN Installation by the Phone Company

The simplest case for ISDN installation will occur if you live within 18,000 feet of one of the phone company's ISDN switches, if you have only a single POTS number at your house, and if your local cable bundle has an unused cable pair suitable for ISDN. Just about all the phone company has to do in that case is to get on their computer and digitally "flip a few switches" to connect everything and test it up to your demarc. They will then send an installation person to your house to connect the two wires of your existing cable to the proper screw posts in your demarc and test the ISDN line with their really cool ISDN line tester. This is all they are really required to do to "install" your ISDN line.

However, if you have your wiring ready, the installer will usually connect it at the demarc and may even go inside and test it for you for FREE. Don't assume anything, though, since many of the RBOCs charge $35 or more just to walk into your house to do anything and then charge $15 every quarter of an hour ($1 per minute) thereafter.

If you have your wiring ready, your computer running, and your SecureLink card setup, you can try using your POTS phone plugged into the SecureLink card to see if the ISDN line works. Before trying this, you should start your computer by turning it completely off and then back on after all of the wiring is completed and plugged in. This insures that the SecureLink looks at your ISDN line and properly runs the RXLOADER and other drivers. If all of them load without error messages, your POTS line should work, and you should be able to use your ISDN line for Internet calls after you configure your Internet software with your ISP's help.

Just a word or two about your friendly phone company ISDN installer. Since ISDN is quite new and predominantly used by businesses, most of the ISDN installers are from the business department rather than the residential department. They are generally knowledgeable and well-educated in dealing with ISDN systems. You'll probably find them extremely polite and quite accommodating if you treat them with respect and in a businesslike manner. They are used to installing larger systems for businesses rather than dealing with hundreds of irate homeowners' noisy phone line problems every time it rains. They really want your ISDN line to work well, and will do everything they can to see that it does.

ISDN TA/NT1 INTERFACE CARD
INSTALLATION AND TESTING

Ahhh! Your SecureLink II card arrived today and you can't wait to plug it into your computer, flip the power switch, and start communicating at 128 Kbps over the Internet. Sorry, but you'll have to wait just a little while longer, because installation and configuration is a nontrivial matter. You understand trivial vs. nontrivial, don't you? When you ask a programmer to turn off the lights in the conference room on his way out, it's a trivial matter. When you ask a programmer to change one number in a program, it may or may not be a trivial matter. When you ask a programmer to add a new feature to the beta version of a program, it's definitely a NONtrivial matter.

Whether you wait for your ISDN line to be installed before you install your SecureLink II card is sort of a catch-22 situation. You can't completely test the ISDN line inside your house without the SL card up and running, and you can't completely test the card without a working ISDN line. You can, however, partially test the SL card without a working ISDN line, so you may as well get as far as you can before the phone company installer arrives.

Unless you are a hardware expert or are born lucky, installation, setup, and checkout of your SecureLink II card will take you anywhere from a few hours to a few days. This is not to say that anything is wrong with the card; rather, installing network cards is complicated by IRQs, I/O base addresses, shared memory addresses, memory managers, DPMS, and network batch files. Working through each of these elements takes considerable effort, especially when you're doing it for the first time.

Since the "Quick Startup Guide" for the SecureLink II takes you step-by-step through the installation and configuration process, and the "Reference Guide" contains examples of files for WFW as well as other networks, we'll quickly take you through the main steps, along with our comments on what the book doesn't tell you.

Note: The step numbers below do not directly correspond to the step numbers in the SecureLink "Quick Startup Guide."

Step 1

This isn't in the SecureLink Guide. Make a backup of your AUTOEXEC.BAT, CONFIG.SYS, WIN.INI, SYSTEM.INI, and PROTOCOL.INI files on another directory of your hard disk as well as on a floppy disk. Several changes will be made to these files during installation and setup, and you want to make sure you can get back to square one if things get too fouled up.

Step 2

Install the SecureLink II card in your computer following ISC's directions.

Step 3

If your ISDN line hasn't been installed, skip the section on plugging the RJ-45 plug of the cable into the U interface on the card. By the way, a standard RJ-11 plug will plug into an RJ-45 jack just fine. Since the ISDN line only uses the middle two wires, in any order it seems, just about any phone wire with RJ-11 plugs on both ends will work with ISDN. To be on the safe side though, you should keep the wire colors the same throughout your wiring—that is, red to red, green to green, etc.

Step 4

Plug your POTS phone into the RJ-11 jack on the card. Of course, it won't work unless you have the ISDN line installed.

Step 5

Turn your computer on. It's the red switch on the side, button on the front, toggle switch on the front, keep looking, you'll find it. :-) .

Step 6

Install a memory manager. You're probably already using EMM386 with your WFW, or you may be using QEMM386, or some other program. The idea is to free up at least 550 kilobytes of conventional memory so that the installation program for the SecureLink II software can run.

It will also help your system's overall performance to use a memory manager. This step may take a while if you're not familiar with DOS and Windows. You can check your system's available memory from DOS by typing MEM /C at the DOS prompt and looking at the number after "Largest executable program size" to see if it is 550,000 or larger. If it isn't, you can run MEMMAKER (if you have DOS 6.*x*) from the DOS prompt (without having Windows running) and let it try to get you some more conventional memory space.

Step 7

After you have freed enough memory, you can install the SecureLink network software. It is a two-step process, but maybe they will integrate it into a single step someday. Just follow the directions in the "Quick Startup Guide."

Phase 2 asks you for the following information about your ISDN service, which you should be able to get from your phone company order line person when they confirm your installation date:

Your ISDN telephone numbers (2)	512 111 0000 and 512 111 0001
Your ISDN SPIDs (2)	512 111 0000-01 and 512 111 0001-01
The telephone company's switch type	National ISDN (NI-1)
Names for your computer and your ISP's computer	You make them up.
Your ISP's ISDN dial-up number.	Put your own ISDN number here for testing

You will also be led through the following SecureLink II card settings:

I/O base address	2A0 - 2BF (hex)
IRQ	11
Shared memory address	CA00 - CBFF (hex)
ISDN interface	U Interface
S/T termination	50 ohm (for U interface)

The first three will be chosen by the installation software, after it looks at your system; however, the program can't really detect all possible conflicts, so these are just "educated" guesses. Leave them the way they are for now, unless you know that another card in your computer (an internal modem, perhaps) uses the same IRQ. It will be obvious when you run other programs if there is a conflict. If that happens, run the setup program again and try another setting. This is where it becomes time-consuming if things don't work the first time, and you don't know where to start changing settings. There is no great way to tell, either.

Your first choice is to use one of the Windows diagnostic programs such as WinProbe. The second is to use the MSD program in DOS, and to look for potential conflicts. This won't tell you about Windows conflicts, but most conflicts occur at the DOS level anyway. If you need to make changes, write them down and change only one setting at a time, then rerun your system to see if it works properly. Keep trying until you get things working, or until you run out of alternatives.

Step 8

Run the RXSETUP program again and choose its board diagnostics routines to let it find any conflicts. Since it made the settings, it probably won't find any problems at this time. If it does, try changing and testing them, one at a time.

Step 9

This step tests the ISDN line itself and can't be used until you have your ISDN line installed and the phone company says it is working properly. These diagnostics actually test the line all the way to the phone company's switch and back. If you try to run this test without a working ISDN line, you will get a "layer 1 link failure," which means the SecureLink card isn't finding a working line. Layer 1 is the first ISDN layer from your NT1. If the phone company's switch isn't properly configured or your card isn't configured for the switch, you will probably get a "layer 2 link failure."

If your ISDN line is installed and properly configured, and the line diagnostics find no errors, you should be able to use your POTS phone to make calls at this point. Try it, and see if it works.

Step 10

This step leads you through performing self-call testing and setup for remote calling. The instructions are clear in the Guide.

Step 11

Setting up the connection protocol is a tricky step. For this step to work, you must first install the SecureLink card and the TCP/IP-32 network driver in the WFW networks section.

Installing the SecureLink Card and TCP/IP-32 (Wolverine) into Windows for Workgroups

Before you can completely configure the SecureLink II card, you must let WFW know it exists and install the TCP/IP-32 stack as its protocol. To do this, you will first need to download the self-extracting wolvbeta.exe file (688,408 bytes) from ftp.microsoft.com and run it to let it uncompress itself.

You then need to run Windows and use the **Drivers** section of the Windows **Setup/Options/Change Network Settings** window to add your SecureLink II card and install the TCP/IP-32 driver. When you click the **Drivers** button, you will see a window with buttons to **Add Protocol** and **Add Adapter**. The SecureLink II is the adapter and the **TCP/IP-32** is the protocol that you need to add. Add the adapter first, then select it, and add the protocol to it. Click the **Help** button if you need it; it's very comprehensive.

If you have another network card installed in your computer, in addition to the SecureLink II card, you may have unused protocols attached to it. Remove any unused protocols, since WFW can only handle a limited number per card, and only one can be NDIS3. If the SecureLink II is the only network card you are using, you only need to install the TCP/IP-32 protocol for it. Remove any others that WFW may install automatically.

Now you can complete the connection protocol portion of the SecureLink II setup as shown in the "Quick Startup Guide." Follow the instructions very, very carefully and select the PPP option when possible for a standard 64 Kbps 1 B channel. If your ISP has ML/PPP at 128 Kbps, select this option after discussing it with your ISP. The following are the settings in the SecureLink setup routine which worked for the authors (of course the addresses, names, and password must be yours):

Remote PC and Other Options

Remote PC name	Realtime
Phone numbers	[3774141] [ ] [ ] [ ]
Remote LAN type	Ethernet
Compression	On
Dialout	On
NetWare 802.3 conversion	Off
B-Channel Speed	Auto
Max. transmission rate	64 Kbits/sec
Channel expansion level	High 0, low 0
Idle disconnect time-out	1 min

Options for PPP Protocol

PPP security	PAP
Username	Yournamehere
Password	XXXXXXX
Bridge control protocol	Off
IP control protocol	On
IP address	244.76.1.123
IPX control protocol	Off
IPX address	n/a
NetBIOS control protocol	Off
NetBIOS name	n/a

At this point, you are instructed to accept the settings and press ESC to exit the RXSETUP program, which saves the settings.

Step 12

You are now finished configuring your SecureLink II card and WFW. However, there may be some confusion on the modification of your CONFIG.SYS and AUTOEXEC.BAT files and the creation of STARTNET.BAT. The SecureLink "Quick Startup Guide" isn't very clear about what goes where.

What needs to be accomplished is simple, in theory. First, your computer must keep from using those part of its memory that the SecureLink card needs for its scratch RAM buffer space. It does this when you add the RAM and X=####-#### sections to the EMM386 line in your CONFIG.SYS file, as shown in the third line below. The DOS Protected Mode Services Device driver (DPMS.EXE) must be loaded as shown in the 12th line below. Finally the SecureLink RXBUFFER must be loaded as shown in the 14th (last) line below. This CONFIG.SYS file worked for the authors, but we can't guarantee that it will work for you:

1. DEVICE=C:\DOS\HIMEM.SYS
2. rem adds RAM and X=####-#### for SecureLink card
3. DEVICE=C:\DOS\EMM386.EXE NOEMS RAM X=CC00-CDFF
4. BUFFERS=40,0
5. FILES=80
6. DOS=HIGH,UMB
7. LASTDRIVE=Z
8. FCBS=16,0
9. STACKS=9,256
10. SHELL=C:\DOS\COMMAND.COM /p /e:512
11. rem loads DOS Protected Mode Services Device Driver (DPMS.EXE)
12. DEVICE=C:\RX\DRIVERS\DPMS.EXE
13. rem adds SecureLink drivers
14. DEVICEHIGH=C:\RX\DRIVERS\RXBUFFER.SYS /S3FFF

Your CONFIG.SYS file may contain other commands in addition to those shown above. The exact address of the X=####-#### line will be given to you by the RXSETUP program. You can add the above commands to your CONFIG.SYS file with the SYSEDIT program in WFW or NOTEPAD, etc.

Next, the RXLOADER program must run and load the RXPCKT.OVL overlay with the /P parameter and setup a DATA file path. It then must use RXCALL to load the PostDrvLoad. After this, you must run the network part of WFW (NET START), then run Windows. All of this is accomplished in the four command lines shown below.

```
C:\RX\DRIVERS\RXLOADER C:\RX\BIN\RXPCKT.OVL /PC:\RX\DATA %1
```

```
C:\RX\RXCALL /PostDrvLoad
C:\WINDOWS\net start
WIN
```

It would be too simple to put these four lines in your AUTOEXEC.BAT file, and it may not run properly either. When the RXLOADER program runs, it checks the status of your ISDN line. If it finds something wrong, it doesn't load the RXPCKT.OVL. When this happens, the RXCALL program aborts. With neither of these loaded, WFW's NET START can't find the SecureLink drivers, so it doesn't load properly. Finally, WFW may load, but it will give you network error messages.

The way around all of this is to place the first two lines at the end of your AUTOEXEC.BAT file and the last two in another batch file such as STARTWIN.BAT. Using this method, your computer automatically boots and starts everything up through the RX drivers. If RXLOADER doesn't properly load, you will see an error message on your screen. You should try running the two RX command lines again to see if they will work the second try, which they usually do unless your ISDN line is "really dead." Putting just the first two RX lines in another batch file (as in START-NET.BAT) just for this use is a good idea. You're wondering why this is necessary. Just remember that the ISDN system is relatively new and doesn't work perfectly every time, yet. Timing is important, especially during startup. So second tries are needed some of the time.

After you see that the RX drivers have loaded properly, you simply run the STARTWIN.BAT file to complete your startup. This completes your setup and configuration of your SecureLink II card, WFW, and your DOS startup files. Now you are ready to get with your ISP to configure your Internet software and ISDN Internet account.

ISDN AND YOUR ISP

Now that you have your ISDN line installed, configured, and your SecureLink card running smoothly with your WFW system, you're ready to complete your Internet connection. Call your Internet Service Provider (ISP) and get them to activate the ISDN account that you ordered a few days/weeks ago. You'll need several pieces of information from them to enter into your system:

Their ISDN phone number	###-####
Your ISDN account number	###
Your IP address	123.12.1.###
Your Default Gateway	123.12.1.###
Your Subnet mask	255.255.255.###
Their DNS (name server) address	198.3.118.11
Your user name	yourusername
Your password	********

You will need to put these pieces of information into your TCP/IP-32 configuration via WFW so it can perform the "dialup" function that perhaps you were having Trumpet Winsock perform with your SLIP account. To do this, mouse into **Windows Setup/Options/Change Network Settings/Drivers** window. Highlight the **Microsoft TCP/IP-32** line under the **ISC SecureLink Adapter [NDIS2]** line and click the **Setup** button. Fill in the **IP address**, **Subnet Mask**, and **Default Gateway** lines using the addresses your ISP gave you.

Now click the **DNS** button. Fill in the **Host Name** with your computer's name (name it yourself). Fill in the **Domain Name** with your ISP's domain name (e.g., bga.com). Fill in the **DNS search order**'s first line with the DNS address(es) that your ISP gave you. This should be all you need to do in this window, so click the **OK** button.

This actually should be all of the information you need to enter in any window so click **OK** on each window to close it and back out of the configuration. Your changes should be saved and WFW will want to restart your computer to put the changes into effect. Click the button and let it do its thing. It will add the appropriate information to several Windows .INI files. After WFW reloads, you should be ready to connect to the Internet.

Whenever you activate your WWW browser, E-mail application, News Reader, or FTP program, your ISDN connection should be established automatically within less than one second. If you have properly configured all of the necessary components discussed above, you should be able to run all of your existing Internet applications via your new ISDN service and PPP accounts just as you previously did using your POTS line and SLIP account, at a much faster rate. If something doesn't work properly and you can't use your Internet account, refer to Chapter 14 for Troubleshooting assistance and call your ISP's customer support line.

Summary

With all of the complexities of installing and configuring the various ISDN components, don't be surprised if everything doesn't work perfectly the first time you try it. Try to be patient and check your setup parameters carefully when you enter them. Proceed deliberately, step by step. Test after each step. Write everything down as if you were the one writing this book for someone else to follow for their first-time installation.

To help you keep your blood pressure down, don't expect your ISDN service to be up and running as continually and reliably as your tried and true POTS line. Accept it as a fact of life that ISDN service is new to everyone, TPC and you included, and that everyone wants it to be as reliable as possible. But it's newly developed, highly technical, completely computerized, digital transmission that is going to be more susceptible to weather-related problems and other perturbations than POTS lines until all of the "kinks" are worked out of the system. Keep the phone company's repair number handy and give them a call anytime you can't call out on your POTS phone that's plugged into the SecureLink's POTS jack when your computer is running and the SecureLink software is all properly loaded and apparently functioning properly.

Check out Chapter 15 for helpful tips and ideas on making your ISDN service more useful and more reliable once you get it up and running.

Chapter 13

Individual Office ISDN Systems

An individual office ISDN system is quite similar to the home system described in Chapter 12. For the purposes of this chapter, a "small office ISDN system" will be defined by these specifications:

1. You want to install the ISDN system yourself
2. Single PC-type computer running DOS 6.x and Windows 3.1, WFW 3.11, Win95, or OS/2 Warp
3. Twenty-four hour use of POTS line via ISDN
4. Attach multiple POTS phones to the same line (same phone number)
5. Full 128 Kbps throughput with analog phone to ring through one B channel
6. Internet connection at full 128 Kbps
7. Capability of using ISDN phones and other ISDN devices (fax, etc.)
8. Use existing analog modem via ISDN (POTS) line

Given the specifications above and the same basic affirmative answers to the questions in Chapter 11 regarding installing the system yourself, you should proceed in the same manner discussed in the previous chapter. This chapter will not reiterate all of those details, but rather will note differences and provide additional instructions regarding both home and non-residential installations. If you desire more complex ISDN services, we strongly suggest you contract with an ISDN consultant. This should help to ensure that you get the best system for your particular needs, in the optimal time frame, at a reasonable price (you can hope so, anyway).

ORDERING THE ISDN SERVICE FROM YOUR LOCAL ISDN DIALTONE PROVIDER

Call your local telephone company's business order number and get connected to their ISDN order person, if they have one. Tell the operator that you want to order an ISDN BRI 2B+D line installed at your place of business (residence or office)... then wait for the inevitable pause from the phone company person. If he or she doesn't pause, but says, "Certainly sir, may I have your present phone number," you've landed a winner who knows what's going on.

After you two determine that you are located in an ISDN service area, get answers for these questions: (assume you got the following answers)

$150	What is the ISDN installation charge?
$70 flat	What are the monthly charges (basic and usage) for ISDN service?
no	Will your ISDN service be via a repeater?
yes	Will your ISDN service be via a SLC?
Siemens	Which central office ISDN switch do they have? (Siemens, AT&T, etc.)
NI-1	Which protocol does the switch use? (i.e., NI-1 [National ISDN-1], AT&T 5ESS Custom, AT&T G3 PBX, or Northern Telecom DMS-100 Custom, Northern Telecom BCS-34 [PVC-1], or other)
NA	What calling features are available? (e.g., EKTS, CACH)
yes	Does your ISDN provider offer BRI (2B+D) with two SPIDs?
yes	Does your ISDN provider offer voice service on both B channels on the same BRI service with two SPIDs?
same price	What are the prices of 2B - data vs. 2B - 1data/1voice vs. 2b—both voice or data?

Write the answers down so you can refer to them later. To meet your requirements for 128 Kbps with automatic switchover for an incoming voice call, you will need your service to have both B channels provisioned with data capabilities and at least one of them with voice capabilities. Go for both voice, if the cost isn't prohibitive, especially if you want to use an ISDN phone or two on your system.

If you want special telephone features similar to those on business Centrex systems and you plan to use an ISDN phone, discuss this with the phone service representative and order those you desire. The features available depend on the RBOC and the type of switch they are using in your vicinity. Some switches have capabilities very similar to Centrex business lines, but some are quite limited in their functionality.

If you like what you hear, ask when the service can be installed. No matter what you are told, you can't do much about it, so you either order the installation or you don't. When you place your order, have the phone company representative confirm the installation price and monthly service charge for the features you have ordered. You will probably be given an order confirmation number and a phone number to dial to check on the progress of your order. Write these down also. You will probably need them. (We're not pessimistic, just realistic.)

While TPC (the phone company) is getting ready to install your line, or turn it on, you can get ready for the big day by purchasing your equipment, software, and signing up for an ISDN account at your ISP.

Purchasing Your ISDN TA/NT1 Interface Card

To meet your specifications for 128 Kbps for Internet, 24 hour POTS usage via ISDN, and the capability of having multiple ISDN devices connected to your system, you will need to make sure that the NT1 and TA or the combined TA/NT1 you purchase meet the following criteria:

- NT1 with separate power supply or combined TA/NT1 with an external power cable, or you are willing to leave your computer turned on 24 hours a day.
- The TA is capable of 2B channel BONDing with automatic switch to voice for an incoming POTS call.

- The TA has at least one and preferably two S/T jacks for ISDN devices.
- Either the NT1 or the combined TA/NT1 has at least one POTS jack.

Several of the NT1s and TAs described in earlier chapters meet these criteria. Talk with your local hardware vendor and your ISP concerning the best combination for your specific needs. You may be able to get the hardware from a mail-order company such as the ISDN Warehouse, or another one you may find via the Internet or in your local area.

Ask around, but remember that it will probably take TPC upwards of three weeks to install your line. If you have to wait for delivery, go on to the next step below. If you're lucky and find the card locally, buy it, take it home, take it out of the box, look at it, put it back in the box, then give your ISP a ring to make sure you can get an ISDN dial-up account using PPP (as their information on their WWW site promises).

Ordering Your ISDN Internet Service

When you call your ISP, confirm that they offer 2B channel BONDing so you can get the 128 Kbps transmission that you specified. Check the cost. Most ISP's will charge you for each 64 Kbps channel you use. The charges may include a monthly service fee in addition to actual per second usage on each channel. As you undoubtedly do with all of your business relationships, get all prices, terms and conditions in writing from your ISP.

Bear in mind that few ISPs have automated the ISDN setup on their end as they have for SLIP accounts. It may take as much as 30 minutes for their network person to configure an account for you. They will want to know what TA and NT1 you are using, in addition to what software you will use to dial-up and connect your PPP link. They will probably tell you that they can be ready for you in a couple of days, which is sooner than you need them to be. Tell them your installation date for your ISDN service, and tell them you will call them back when it gets installed and you get your card installed in your computer.

ISDN DIALTONE SERVICE: WIRING, INSTALLATION, CONFIGURATION, AND TESTING

Wiring Your Office for ISDN

If you're installing the ISDN in your home office, follow the instructions in Chapter 12. If your location is a separate business office, you may need to contact the owner or property manager before doing any wiring inside your office. Since most office buildings have telephone wiring panels inside closets on either each floor or sometimes in each office suite, wiring shouldn't be a problem. The phone company ISDN installation person will bring the ISDN line at least to the nearest wiring panel and perhaps all the way to your office. Most offices are wired with at least 8-wire twisted-pair cable, which should give you plenty of unused pairs, only one of which is needed for your ISDN service.

Since this is a business installation, you may need to call in a phone wiring company if your office property manager or owner requires it. If they do not require it, and if you are handy with a screwdriver, you can easily plug the wires into the proper spots on the wiring panel, *if* you can get to it, and *if* you know which wires are yours. It may be most cost-effective to have the phone company installation person go ahead and plug in the wires all the way to your office and test them at that location. The installation person can also put an RJ-45 jack on the end of the cable if you don't already have one in the wall of your office. Most multi-line business telephones use RJ-45 plugs and jacks, so many offices already have the proper jacks in the wall.

For a comprehensive look at ISDN wiring, contact the North American ISDN Users' Forum (NIUF) at 301-975-2937 or e-mail *dawn@isdn.ncsl.nist.gov* and request the "ISDN Wiring Guide for Residential and Small Businesses." It will tell you much more than you probably care to know, but who knows, it may make for interesting conversation at your next ISDN users' group meeting.

ISDN Installation by the Phone Company

The simplest case for ISDN installation will occur if your business location is within 18,000 feet of one of the phone company's ISDN switches. Just about all the phone company has to do in this case is get on their computer and digitally "flip a few switches" to connect everything and test it up to your office. They will then send an installation person to your location to connect two wires of your existing cable to the proper punch-down tabs in your wiring panel and test the ISDN line with their really cool ISDN line tester. This is all they are really required to do to "install" your ISDN line.

If you have your computer running, and your NT1 and TA all set up, you can try using your POTS phone plugged into the TA or NT1 to see if the ISDN line works. Before trying this, you should start your computer by turning it completely off and then back on after all of the wiring is completed and plugged in. This insures that the TA looks at your ISDN line and properly loads the drivers. If all of them load without error messages, your POTS line should work, and you should be able to use your ISDN line for Internet calls after you configure your Internet software with your ISP's help.

ISDN TA and NT1 Installation and Testing

Ahhh! Your NT1 and TA arrived today, and you can't wait to plug it into your computer, flip the power switch, and start communicating at 128 Kbps over the Internet. Sorry, but you will have to wait just a little while longer; installation and configuration is more complicated than programming your VCR, but not by much.

If you purchased a combined TA/NT1 card or an internal TA card, the installation should be somewhat similar to the installation discussed in Chapter 12. Just follow the instructions in the card's User's Guide and you will probably be finished in only a few hours. Remember these cards are actually network cards and require all of the setup that any network card requires.

If you are using a standalone TA unit that plugs into your existing Ethernet card, the setup may be considerably less complex. Again, carefully follow the TA's instructions for installation and setup. Many of the steps will be the same as for a card, but without most of the network configuration, since it will be acting as an independent device on the network.

We will presume you aren't using a serial TA device, since none of them are actually capable of full 128 Kbps transmission rates in uncompressed mode because of the bottleneck of even a fast UART in your computer.

If you are using a separate NT1, set it up according to the manufacturer's instructions. You can quickly test your ISDN line via the POTS jack on the NT1 and your POTS phone when you have the NT1 properly configured.

You can't completely test the ISDN line at your location without an NT1 up and running, and you can't completely test your TA or NT1 without a working ISDN line. You can, however, partially test the TA without a working ISDN line, so you may as well get as far as you can before the phone company installer arrives.

Remember, always make a backup of your AUTOEXEC.BAT, CONFIG.SYS, WIN.INI, SYSTEM.INI, and PROTOCOL.INI files on another directory of your hard disk, as well as on a floppy disk prior to starting the installation and setup of any TA device. Several changes will be made in these files during installation and setup, and you want to make sure you can get back to square one if things get too fouled up.

When setting up your NT1 and your TA, you will need all or most of the following information at one time or the other. So write it down in a list and keep it handy. Some of the parameters shown concern other types of networks that you may have installed on your computer if your office has a network. This chapter isn't really about networked computers, but we've included these parameters to show you that they exist and that they can be included in your setup if you need them.

Your ISDN telephone numbers (2)	512 111 0000 and 512 111 0001
Your ISDN SPIDs (2)	512 111 0000-01 and 512 111 0001-01
The telephone company's switch type	National ISDN (NI-1)
Names for your computer and your ISP's computer	You make them up.
Your ISP's ISDN dial-up number.	Put your own ISDN number here for testing

Remote PC and Other Options

Remote PC name	Realtime
Phone numbers	[3774141] [ ] [ ] [ ]
Remote LAN type	Ethernet
Compression	On
Dialout	On
NetWare 802.3 conversion	Off
B-Channel Speed	Auto
Max. transmission rate	128 Kbits/sec
Channel expansion level	High 0, low 0
Idle disconnect time-out	1 min

Options for PPP Protocol

PPP security	PAP
Username	Yournamehere
Password	XXXXXXX
Bridge control protocol	Off
IP control protocol	On
IP address	244.76.1.123
IPX control protocol	Off
IPX address	n/a
NetBIOS control protocol	Off
NetBIOS name	n/a

Depending on your specific TA, this will probably complete its installation and configuration. Your chosen TA will have at least one S/T jack on it to provide access for either an ISDN telephone or another ISDN-capable device should you desire to use one. Although it's early in the game, there are a few ISDN fax machines available, and there may be some ISDN answering machines available in the near future.

Although you can plug your standard fax, answering machine, or analog modem into the POTS jack on your NT1 or your TA/NT1 card, be aware that you may be able to transmit out via the fax or modem, but you may not be able to automatically receive information. This is caused by the lack of the analog ring voltage being passed through the NT1 or TA/NT1 card to your analog device, which requires the ring voltage before it will "answer" the call.

Although most TA card manufacturers say they don't support it, you can plug your internal modem into the POTS jack and call out on one of your ISDN channels even while you are using the other channel for data. Most of the ISDN cards will work in this manner. The manufacturers are hesitant to say they "support" this type of usage because problems that occur when using this combination of analog modem and digital ISDN card can be very difficult to solve. They don't want the hassle and added expense incurred if they advertise that they "support" this use.

Installing TCP/IP

Before you can completely configure your TA, you must let Windows, WFW, Win95, or OS/2 Warp know it exists. You must also install a TCP/IP stack such as TCP/IP-32 for WFW 3.11 or another for Windows 3.1. Both Win95 and OS/2 Warp contain their own software that provides the functionality of a TCP/IP stack, so refer to their manuals for its configuration.

If you are using WFW 3.11, please refer to Chapter 12 for installation of the TCP/IP-32 stack and its setup. Otherwise, obtain the appropriate protocol driver for your TA and install it according to its directions. Your ISP may be able to help you if you have been in contact with them regarding your choice of hardware and software. It is in their best interest to see that you can easily connect to their system so they can collect your money for using their system.

No matter what type of software you choose, take your time and follow the instructions very, very carefully.

ISDN and Your ISP

Now that you have your ISDN line installed, configured, and your NT1 and TA running smoothly with your computer system, you're ready to complete your Internet connection. Call your Internet Service Provider (ISP) and get them to activate the ISDN account that you ordered a few days/weeks ago. You'll need several pieces of information from them to enter into your system:

Their ISDN phone number	###-####
Your ISDN account number	###
Your IP address	123.12.1.###
Your default gateway	123.12.1.###
Your Subnet mask	255.255.255.###
Their DNS (name server) address	198.3.118.11
Your user name	yourusername
Your password	********

You will need to put these pieces of information into your TCP/IP or other similar program's configuration so it can perform the "dial-up" function that perhaps you were having Trumpet Winsock perform with your SLIP account. If you are using WFW 3.11 and TCP/IP-32, follow the instructions in Chapter 12. Otherwise, follow the instructions in your chosen software's User's Manual. After you have configured your program, you should be ready to connect to the Internet.

Whenever you activate your WWW browser, E-mail application, news reader, or FTP program, your ISDN connection should be established automatically within less than one second. If you have properly configured all of the necessary components discussed above, you should be able to run all of your existing Internet applications via your new ISDN service and PPP accounts just as you previously did using your POTS line and SLIP account, at a much faster rate. If something doesn't work properly and you can't use your Internet account, refer to Chapter 14 for troubleshooting assistance and call your ISP's customer support line.

SUMMARY

You're a business person. Therefore, you know that nothing works perfectly the first time you try it. Murphy's law just won't let that happen. So try to be patient and check your setup parameters carefully when you enter them. Proceed deliberately, step by step. Test after each step. Write everything down as if you were the one writing this book for someone else to follow for their first-time installation.

If you have problems, Chapter 14 may help you. Also, call your ISP if the problems are in connecting to their Internet system. If your POTS phone won't work at all, give the phone company a call. They can use their computers to check your ISDN line fairly far up toward your location without sending anyone out. Also, check out Chapter 15 for helpful tips and ideas on making your ISDN service more useful and more reliable once you get it up and running.

Chapter 14
Troubleshooting, Tips, and Testing

Troubleshooting Your ISDN System

ISDN Line Problems

ISDN lines fail differently from analog lines. With an analog line you can get poor voice reception, noisy lines and even crossover to other conversations. With an ISDN line, you may occasionally get a "pop" or "snap" on the line, but when an ISDN line fails, it usually won't work at all.

If you're not sure if your ISDN line is down, or your NT1 or TA isn't working properly, try plugging your POTS phone into the ISDN jack. If you hear "white noise," the ISDN line is still there, in some fashion. If it is dead calm, the ISDN line is completely out and it's time to call your local phone company's repair number.

If your ISDN drivers try to load and give you a Layer 1 Failure error, there is probably something wrong with the ISDN line itself. If you get a Layer 2 Failure error, the problem is probably with the provisioning of the phone company's switch for your service.

In testing your TA, you may be instructed to perform a "self call" test. Some instructions have you call your secondary ISDN number via your primary number. For some reason, this may not work, although if you try to call your primary ISDN number, it will work. Try both and see what happens. This is probably a function of the way the ISDN service is provisioned for data. Call your ISDN provider if neither works and you can't get any data call to function.

Most TA software packages include a "logging" function which is useful for tracking down problems. The log files keep a running account of what happens during your ISDN sessions. They generally contain the following type of information (as shown in this example taken from our ISC SecureLink log):

```
06/26/95 11:18:45 *** Logging STARTED ***
06/26/95 11:18:45 SecureLink(TM) ISDN Adapter, Version 3.00b Feb 09 1995
06/26/95 11:18:45 TAPI Version 3.00b Feb 09 1995
06/26/95 11:18:45 Switch type: National ISDN-1
06/26/95 11:18:45 Board 1:
06/26/95 11:18:45   S/T interface, iobase = $2a0, IRQ = 11, address = $cc00
06/26/95 11:18:45    SPID #1 = 512123123401
06/26/95 11:18:45    SPID #2 = 512123123501
06/26/95 11:18:46 Physical layer connection established for board 1
06/26/95 11:18:46 Layer 2 connection established for board 1, TEI 105
06/26/95 11:18:46 Layer 2 connection established for board 1, TEI 106
06/26/95 11:18:46 SPID #1 accepted for board 1
06/26/95 11:18:46 SPID #2 accepted for board 1
06/26/95 11:18:46 Initialization successful.
06/26/95 11:18:58 Attempting to establish PPP connection to REALTIME.
06/26/95 11:19:01 Call established (64 Kbps) to REALTIME at 3774141.
06/26/95 11:19:01 PPP - Authentication Complete to REALTIME.
06/26/95 11:19:01 PPP - IP Link UP to REALTIME (213.88.3.133)
06/26/95 11:19:01        Local IP Address (213.88.3.124) was configured.
06/26/95 11:20:03 Call down (64 Kbps) to REALTIME at 3774141.
06/26/95 11:20:03      Reason -> NORMAL.
06/26/95 11:20:03 PPP connection terminated to REALTIME.
06/26/95 12:18:03 Attempting to establish PPP connection to REALTIME.
06/26/95 12:18:04 Call established (64 Kbps) to REALTIME at 3774141.
06/26/95 12:18:04 PPP - Authentication Complete to REALTIME.
06/26/95 12:18:04 PPP - IP Link UP to REALTIME (213.88.3.133)
06/26/95 12:18:04        Local IP Address (213.88.3.124) was configured.
06/26/95 12:19:07 Call down (64 Kbps) to REALTIME at 3774141.
06/26/95 12:19:07      Reason -> NORMAL.
06/26/95 12:19:07 PPP connection terminated to REALTIME.
```

The log includes any error messages that occur during operation. These should be defined—with suggestions on how to remedy the problem—in your card's users manual. You can always give the vendor or manufacturer or your ISDN TA a call on your working POTS line to help you diagnose and repair your problem.

IRQ, Address, SRAM, and Other Hardware Conflicts

Newer PCs have various built-in system diagnostics and settings that are designed to provide faster, smoother operation. But when you start adding complicated network cards and drivers, some conflicts invariably do occur. The standard IRQs, I/O addresses, and SRAM (scratch RAM) conflicts are usually noticed by the installation program from the ISDN card. Alternatively, you can use the Microsoft Diagnostic program (MSD.EXE) or a Windows-based diagnostic program to determine your current hardware and software settings and usage, and then set the card to avoid conflicts.

Conflicts with the PC's BIOS and CMOS settings are much harder to find. It generally requires using a process of elimination by testing one setting at a time, until the conflict is located (or fixed). Since this requires that you reboot your computer each time you try a different setting, it is very time-consuming.

Comments from users of various systems have pointed to the following known conflicts, so you might start by disabling the settings if they are present on your computer:

- Auto Config Function—AMIBIOS ADVANCED CHIPSET SETUP
- Hidden Refresh—AMIBIOS ADVANCED CHIPSET SETUP

These two functions, when enabled, kept WFW from loading properly with the ISC SecureLink card and NDIS drivers. When disabled, the system worked just fine with no noticeable degradation in functionality, which makes you wonder what enabling the two "features" really did in the first place.

ROM shadowing in a Compaq ProSignia seemed to cause memory conflict problems when the drivers were being installed for a Digi PCIMAC/4 board under the Windows NT Advanced Server.

If you don't know how to check your CMOS settings, give your hardware vendor a call, and ask for help. They deal with these problems every time they change a modem or another I/O card in a PC. If you think you know enough to try to fix your system yourself, remember to write everything down on screen prints of the settings, to change only one setting at a time, and to cold-boot your computer after each change. Entering **Ctrl+Alt+Delete** may not reset the hardware, and pushing the reset button may not completely clear the RAM in some computers. When in doubt, turn your machine off for at least 10 seconds before restarting it.

Windows for Workgroups and Windows NT Problems and Solutions

If you installed a network ISDN TA/NT1 card, you may notice that your network occasionally creates spurious, periodic ISDN connections to your ISP for no apparent reason (we mean here that you'll observe a connection being established every one to five minutes). If your minimum time-out is one minute and your system automatically reconnects one minute after it has been disconnected, you will be "on-line" for about 30 out of every 60 minutes. This adds rapidly to the charges from both your ISDN and Internet providers.

Several causes, and a couple of solutions, have been suggested for this problem. On systems using Windows NT, the "browser" is looking for other NT machines in your workgroup or domain. Since you apparently have an ISDN interface that looks like an NDIS interface, when it sends local subnet broadcasts your ISDN tries to call the other end of the line to pass that broadcast along.

The solution is easy, because NT has a neat configuration screen in the Network Control Panel. Start Control Panel, and open the Network panel. Then click on "Bindings." On this screen, items on the left represent services, protocols, etc. Items on the right represent interfaces and lower-level protocols. If the light bulb is on, the software pieces are talking to each other. If the bulb is dim, they aren't. You want to turn all the bulbs other than TCP/IP off that lead to your ISDN interface. Just figure out which ones these are and double-click them to do so.

On WFW systems, the solution isn't as simple. The only solution one user has found is to use the Network Settings section of the Windows Setup program to turn off sharing for both files and printers on the ISDN-equipped PC. This stops spurious connections, but it also keeps other PCs from sharing files and printers, which is why you installed WFW in the first place. Maybe someone will come up with a better solution; otherwise we have to live with it until we switch to Win95 (which we can hope won't have the same problem). So far, nobody has brought this problem up on the newsgroups for Win95.

Tips for Better ISDN Use

Would ISDN work to replace my house's POTS service?

If one person is using an ISDN phone and someone in another room picks up an ISDN phone in another room and attempts to access the same B channel, will this work? How is the whole family supposed to talk to Grandma on an ISDN line? Use a speakerphone? Crowd around the same handset?

These are some of the primary reasons that ISDN isn't recommended for general home use. If you really wanted to just use your ISDN BRI service for all of your home needs, you could install a TA or an NT1 with a POTS jack. Then you'd plug the existing house phone wiring into one of the POTS jacks (instead of into the demarc outside your house), thereby keeping your house phone system analog up to the TA or NT1. As long as your TA or NT1 remains powered up, your home analog phones still work the same way as before, except you can call them with one of your ISDN line numbers and they won't ring. The ring voltage isn't usually passed on from the TA or NT1, which is yet another reason not to replace your home POTS with ISDN.

If you're using only ISDN telephones in your house, with the proper wiring and necessary number of S/T jacks on TAs, then ISDN allows for multiple directory numbers and multiple call appearances for each number. With a keyset feature option, the same number can appear on both phones, but for both to be on the *same* call at once, a conference bridge is required (this is automatic with the feature). Also note that it requires two SPIDs, and is usually implemented with two lines, though two SPIDs can share a line and support a conference bridge (at least with AT&T; probably with NI, too).

When you "bridge on" to the call by pressing the lighted button, the CO switch automatically picks out a conference bridge. This is required because ISDN sets are digital. On an analog line, you can put two phones in parallel and their microphone audio signals (analog) will add together, creating a composite. On a digital line, that wouldn't work—adding 1's and 0's together would create random noise at best. So the switch "magically" inserts a bridge when you press a common line appearance on two sets.

How can I most efficiently use my current analog modem with my ISDN service?

You can plug your current analog modem into the POTS jack of your NT1 or your TA and use it exactly as you currently use it for outgoing calls. If you have a fax/modem, you can use it the same way, too. But the receipt of calls changes with some NT1s and TAs: Most fax/modem cards or standalone devices rely on the analog ring voltage to initiate their answering responses. Because some of the NT1s and many of the combined TA/NT1 cards don't pass the analog ring voltage (about 110 volts) through to the analog circuit, this keeps the fax or modem from knowing that there is a phone ringing for it.

If you only receive faxes after someone calls you on your voice line to tell you it's on the way, you can always manually run your fax software and click the manual receive button when you hear your TA card "ring." If you really need unattended fax receipt using your ISDN line, you should either purchase an NT1 that passes the ring voltage through to your analog fax machine, or purchase an ISDN fax device that plugs into the S/T jack on your TA.

How can I keep my Internet connect time to a minimum?

Since your ISDN service can connect in just a second or three, you don't need to be continually logged into your Internet account. Set the time-out on your ISDN card's software to a short time (many offer one-minute increments only, with one minute the shortest allowable time-out). If you're being charged by the minute, this doesn't pose a problem, but if you're being charged by the second, you might try to find software with a shorter time-out, or try to get your software provider to reduce its minimum settings in their next upgrade. A time-out setting of 10 seconds is reasonable for an ISDN line.

Using an off-line news reader can also greatly reduce your connect time if you subscribe to many newsgroups. The FreeAgent news reader from FortÈ (*http://www.forteinc.com*) quickly downloads article headers from the newsgroups you select. You can scroll through them while it is still downloading others; it can also download the bodies of selected articles while you are reading others as well. FreeAgent keeps a database of headers and articles on your hard disk so that you can read them at your leisure after you have disconnected from the ISDN provider, thereby stopping both your ISDN and your ISP charges.

Download large files during low-traffic times on the Internet. Because your ISDN line provides such fast transfer speed, your overall transfer rate may be slowed by bottlenecks on the Internet itself, or on the server from which you are retrieving the file. Try to determine an off-peak time of the day or night for that server (don't forget about time differences, either) and proceed accordingly.

You can always turn your Web browser's image display off, too. Of course, images do appear more quickly via your ISDN line, but copying and rendering images still takes time. If you want to see them you can always just click the images button (or whatever similar command your browser requires) and they will be quickly transferred and displayed. Turning off image display can cut your Internet connect time by as much as a third, so unless you're working on a topic that absolutely requires images to be displayed, leaving them turned off can save you money!

ISDN Line Testing Equipment

While you may desire your very own BRI analyzer, the truth is you will probably never really need one. If you suspect your line isn't working properly, a quick call to your phone company's repair line will get it checked and fixed right away in most cases, and at a much lower price, if any, than the lowest-priced BRI analyzer. The only BRI analyzer that costs under $3000 that we've found is the PA 100 protocol analyzer at $2495. Until it arrived, ISDN protocol analyzers were typically priced from $8000 to over $ 35000. This made it difficult to justify their purchase, especially if you just needed one occasionally.

Without a protocol analyzer, it is almost impossible to resolve the "finger pointing" that commonly occurs between the service provider and the equipment vendor or the customer. If you are in a company that's plagued by ISDN problems, a knowledgeable technician with a protocol analyzer may be able to resolve these problems quickly and at a lower cost than the opportunity costs for the time and productivity lost because of an inoperative ISDN line.

The UPA 100 is a compact unit (9.5" x 7.5" x 1.5") that connects to the serial port of an IBM-PC compatible or Mac computer. It captures the messages passing between the subscriber's equipment and the network, decodes them into plain English, displays them on-screen, and places them in a buffer for later analysis. Circuit-switched and packet-switched decodes for National ISDN, Northern Telecom DMS 100 Custom, and AT&T #5ESS ISDN are supported. When you need a piece of equipment like this, nothing else will do (so you might want to see if any service companies in your area can provide one with a technician for an hourly rate, or check to see if a short-term rental or lease might be available).

Summary

The best advice for keeping your ISDN problems to a minimum, solving the ones you do have quickly, and maximizing the use of your ISDN service can be stated quite simply: "Read the comp.dcom.isdn newsgroup regularly." Ask for help from your ISP, your ISDN service provider, your ISDN hardware vendor, and the folks on the comp.dcom.isdn newsgroup.

Remember that nobody can help you diagnose a problem that you can't explain in adequate detail. Nobody will want to even try to help you if you tell them, "Yes, there was an error message on the screen but I don't remember what it said and I didn't write it down." Write everything down, including the circumstances under which the problem occurred, your hardware and software settings and configuration, and anything else you can think of that might possibly affect the situation. Let the experts ignore the parts that aren't germane to the problem. Good luck!

Frequently Asked Questions and Answers

chapter 15

The questions and answers in this chapter have been excerpted primarily from the comp.dcom.isdn newsgroup and its FAQ. The authors have edited the material to make it more useful to novice ISDN users. Our thanks to all of the subscribers to comp.dcom.isdn for their great questions and cogent answers. Special thanks are needed for a few of the professionals who always seem to be available to patiently answer the questions, day after day, with great skill and forbearance: Chuck Sederholm, IBM Corp.; Laurence V. Marks, IBM Corp.; Fred R. Goldstein, Bolt Beranek & Newman Inc.; and Pat Coghlan, Newbridge.

Chapter Contents

What will Basic Rate (2B+D) ISDN look like in my house/office?

Can I put more than two ISDN devices on my BRI line?

Is it possible to connect more than one NT1 to the U point/ISDN 2 wire line?

Are there any ISDN BBSs in the U.S.?

Can my existing analog telephone lines be used for ISDN?

How does ISDN compare to regular (analog) telephone lines?

Is caller ID available on ISDN?

What do ISDN phones cost?

What is National ISDN?

What is the NIUF?

What is ATM?

What is B-ISDN?

What is BONDing?

Is full-motion video offered over ISDN?

What is a SPID? How come my ISDN device won't work without one?

Can I purchase European ISDN devices and use them successfully in the U.S.?

Do different manufacturers' terminal adapters interoperate when used asynchronously?

How long should call setup take when using a TA?

What are the differences between a bonded 128 Kb ISDN connection and a switched 56 line?

How would ISDN work if I replaced my house's POTS service with it?

What will Basic Rate (2B+D) ISDN look like in my house/office?

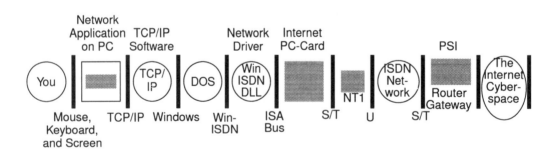

Figure 15.1 *Layers between you and the Internet*

Can I put more than 2 ISDN devices on my BRI line?

Q: Is it possible to put three ISDN phones (using the S/T bus) on one BRI ISDN line? Ameritech states that there is a limit of two devices that can use the same B channel. It's true that you can't use more than two B channels at one time, but what doesn't make sense, since we don't plan to use more than one or two phones at the same time. Why it isn't possible to have up to eight ISDN phones installed? Don't they just monitor the D-channel for signaling information, and only use up one B channel when you pick u the phone? By the way, the ISDN line in question is provisioned by an AT&T 5ESS switch, if that makes any difference.

A: Your problem is not a technical one, it is a policy issue. Your technical assumptions are entirely correct, and you should be able to have eight phones on an AT&T 5E provided BRI.

Your provider may have set a policy that they will not provide a service to a customer served by one switch type, when that service is not available to all switch types. Therefore, since a DMS 100 is not capable of more than two devices per B channel, they will not offer more than two on any switch type.

Their logic is that this will prevent someone served by a DMS100 from demanding an FX'd line off a 5E switch. Also, we understand that DMS100s are much cheaper, and they don't want people to get used to the superior 5E. Hopefully this policy will go away when (and if) NI2 becomes available on the DMS (but we wouldn't bet on it). We also imagine they hope people will buy more lines.

Is it possible to connect more than one NT1 to the U point/ISDN 2 wire line?

NO. The U loop terminates a single NT1 (Network Termination) device per ISDN U line. Therefore, if the NT1 functionality is built into the ISDN equipment, a telephone set, or a data terminal adapter, no additional NT1s can be attached to that ISDN U line. The ISDN equipment with the integrated NT1 would need to have an S/T bus output connection to allow additional ISDN equipment to be attached via an S/T interface.

As a product example, IBMs TE 7845 provides integrated NT1 functionality together with an analog POTS connection which can use one of the ISDN B channels for voice. The second B channel is then made available via an S/T bus connector so an additional ISDN S/T bus device can be connected. In general, the only way to support multiple ISDN devices on a single ISDN U line is by using an NT1 and then connecting the multiple ISDN equipment at the S/T bus connector.

Are there any ISDN BBSs in the U.S.?

The NORTEL Digital Velocity BBS has an ISDN BBS list. You can access this ISDN BBS at *http://www.isdn.nortel.net* and Telnet in, or make an ISDN call to (919) 992-0407 for ISDN access up to 115.2 Kbps using the V.120 communications protocol.

Can my existing analog telephone lines be used for ISDN?

According to Bellcore, usually yes. Most of the analog lines currently in service do not require any special conditioning. However, if a line has load coils or bridge taps installed, your telephone company installation person will be able to "decondition" the line for ISDN use, usually without your knowledge or intervention. In North America, around 90% of existing telephone lines need no "deconditioning" in order to be used for ISDN BRI service.

How does ISDN compare to regular (analog) telephone lines?

A "single" ISDN BRI line may act like two independent analog phone lines with two numbers and be capable of handling data transmission at 64 Kbps per line (B channel). Depending on the central office equipment, many "special" features may be available (conferencing in the telephone switch). BRI ISDN phones can support key-set features like those you would expect to get on an office PBX:

Multiple directory numbers per line	Speed call
Multiple lines per directory number	Call park
Conferencing features	Call pickup
Forwarding features	Ring again
Voice mail features	Status displays

Is Caller ID available on ISDN?

Caller ID (name or number display) may be supported (depending on the CO setup). The availability of Caller ID for residential phones would depend on the capabilities of the local phone network and legislation allowing or disallowing caller ID. The availability of Caller ID relies on the underlying switching protocol used by the switches that make up the telephone system. If Caller ID is available to your ISP, it should be used to help secure your account. Having the ISP's system check to see if the ISDN call requesting your account is coming from either of your ISDN phone numbers greatly increases your security. It's practically impossible for someone who has stolen your account name and password to get around the caller ID number check.

What do ISDN phones cost?

The ISDN sets can cost from $180 for an AT&T 8503T ISDN phone from Pacific Bell, up to $1900, depending on what/how many features are needed. A recent report states that the price is $536.90 for an AT&T 7506 with the RS-232 port on the back and $102.70 to get the 507A adapter to hook analog devices to the 7506. Recent quotes were $200 for a Coretelco 1800 and $600 for a Fujitsu SRS 1050.

What is National ISDN?

Because of the breadth of the international ISDN standards, there are a number of implementation choices that vendors of ISDN equipment can make. Given the number of choices available to vendors, different vendors' equipment may not interoperate. In the United States, Bellcore has released a series of specifications to try to avoid such interoperability problems, including the National ISDN specifications. Contact the Bellcore ISDN hot line for more information (see the vendor list for the number).

What is the NIUF?

North American ISDN Users Forum (NIUF) is an organization of ISDN-interested parties, coordinated by NIST (National Institute of Standards. and

Technology.), which promotes the use of ISDN, and provides a great deal of information about ISDN to anyone who asks for it. Contact:

NIUF Secretariat
National Institute of Standards and Technology
Building 223, Room B364
Gaithersberg, MD 20899
Tel: (301) 975-2937
Fax: (301) 926-9675
(301) 869-7281 BBS 8N1 2400 bps

What is ATM?

ATM (Asynchronous Transfer Mode) is a switching/transmission technique where data is transmitted in small, fixed-size cells (5 byte header, 48 byte payload). The cells lend themselves both to the time-division-multiplexing characteristics of the transmission medium, and to the packet-switching characteristics desired from data networks. At each switching node, the ATM header identifies a "virtual path" or "virtual circuit" for which the cell contains data, enabling the switch to forward the cell to the correct next-hop trunk. The "virtual path" is set up through the involved switches when two endpoints wish to communicate. This type of switching can be implemented in hardware, almost essential when trunk speeds range from 45Mbps to 1.2 Gbps.

One use of ATM is to serve as the core technology for a new set of ISDN offerings known as Broadband ISDN (B-ISDN). For more information, read *comp.dcom.cell-relay*. This group has a Frequently Asked Questions list; it is posted to *news.answers* and is in various archives as "cell-relay-faq."

What is B-ISDN?

Broadband ISDN refers to services that require channel rates greater than a single primary rate channel. While this does not specifically imply any particular technology, ATM will be used as the switching infrastructure for B-ISDN services.

B-ISDN services are categorized as follows:

Interactive:

Conversational—such as videotelephony, videoconferencing
Messaging—such as electronic mail for images, video, graphics
Retrieval—such as teleshopping, news retrieval, remote education

Distribution:

Without user presentation control—electronic newspaper, TV distribution

With user presentation control—remote education, teleadvertising, news retrieval

What is BONDing?

BONDing is a set of protocols developed by the U.S. inverse multiplexer that supports communication over a set of separate channels as if their bandwidth were combined into a single coherent channel. For example, BONDing supports a single 128 Kbps data stream over two 64 Kbps channels.

The specification defines a way of calculating relative delays between multiple network channels and ordering data such that what goes in one end comes out the other. Most vendors also have their own proprietary methods that usually add features and functions not present in BONDing mode 1. Mode 1 is the mode used for recent interoperability testing between vendors.

Is full-motion video offered over ISDN?

In ISDN, video isn't a "service being offered"--at least not for low to midrange quality. To get full-motion video via ISDN, just buy the proper equipment for both subscribers, plug it in, and place the call yourself.

Video telephony over narrowband ISDN is governed by a suite of ITU-T (formerly CCITT) interoperability standards. The overall video telephony suite is known informally as p * 64 (and pronounced "p star 64"), and formally as standard H.320. H.320 is an "umbrella" standard; it specifies H.261 for video compression, H.221, H.230, and H.242 for communications, control, and indication, G.711, G.722, and G.728 for audio signals, and several others for specialized purposes. A common misconception, exploited by some equipment manufacturers, is that compliance with H.261 (the video compression standard) is enough to guarantee interoperability.

Bandwidth can be divided up among video, voice, and data in a bewildering variety of ways. Typically, 56 Kbps might be allocated to voice, with 1.6 Kbps for signaling (control and indication signals), and the balance allocated to video.

An H.320-compatible terminal can support audio and video in one B channel using G.728 audio at 16 Kbps. For a 64 Kbps channel, this leaves 46.4 Kbps for video (after subtracting 1.6 Kbps for H.221 framing).

The resolution of a H.261 video image is either 352×288 (known as CIF) or 176×144 (known as quarter-CIF or QCIF). The frame rate can be anything from 30 frames/second on down. Configurations typically use a 2B (BRI) or a 6B (switched-384 or 3xBRI with an inverse multiplexer) service, depending on the desired cost and video quality. In a 384 Kbps call, a video conferencing system can achieve 30 frames/second at CIF, and looks comparable to a VHS videotape picture. In a 2B BRI call, a standard video phone can achieve 15 frames/second at CIF.

Those who have seen the 1B video call in operation generally agree that the quality is not sufficient for anything useful like computer-based training—only for the social aspect of being able to *see* Grandma as well as hear her (sort of like the snapshot pictures you make with that $5 camera with no controls).

A 2B picture, on the other hand, is for all practical purposes sufficient for remote education, presentations, etc. Rapidly changing scenes are still not very well handled, but as soon as the picture calms down, the sharpness and color quality are impressive (considering that only two plain phone channels are being used). With 2B+D as the standard BRI, this kind of picturephone will be usable everywhere (including private homes).

However, it should still be noted that 6xB or H0 does allow for dramatic improvement in picture quality compared to 2xB. In particular, H.320 video/audio applications will often allocate 56 Kbps for audio, leaving only 68.8 Kbps for video when using 2xB. On the other hand, using H0 would get you 326.4 Kbps for video with 56 Kbps for audio. Alternative audio algorithms can improve picture quality over 2xB by not stealing as many bits. Note that 6B is not identical to H0; the latter is a single channel which will give you 80 Kbps above that of six separate B channels. Inverse multiplexers can be used to combine B channels.

What is a SPID? Why won't my ISDN device work without one?

SPIDs are Service Profile IDs. SPIDs are used to identify what sort of services and features the switch provides to the ISDN device. Currently they are used only for circuit-switched service (as opposed to packet-switched). Annex A to ITU recommendation Q.932 specifies the (optional) procedures for SPIDs. They are most commonly implemented by ISDN equipment used in North America.

When a new subscriber is added, the telco personnel allocate a SPID just as they allocate a directory number. In many cases, the SPID number is identical to the (full 10-digit) directory number. In other cases it may be the directory number concatenated with various other strings of digits, such as digits 0100 or 0010, 1 or 2 (indicating the first or second B channel on a non-Centrex line), or 100 or 200 (same idea but on a Centrex line) or some other, seemingly arbitrary string. Some people report SPIDs of the form 01$nnnnnnn$0 for AT&T custom and 01$nnnnnnn$011 for NI-1, where n is the seven-digit directory number. It is all quite implementation-dependent.

You need to configure the SPID into your terminal (i.e., computer or telephone, etc., not your NT1 or NT2) before you will be able to connect to the central office switch. When you plug in a properly configured device to the line, Layer 2 initialization takes place, establishing the basic transport mechanism. However, if you have not configured the given SPID into your ISDN device, the device will not perform layer 3 initialization and you will not be able to make calls. This is, unfortunately, how many subscribers discover they need a SPID.

Once the SPID is configured, the terminals go through an initialization/identification state, which has the terminal send the SPID to the network in a Layer 3 INFOrmation message whereby the network responds with an INFO message with the EID information element (ie). Thereafter the SPID is not sent again to the switch. The switch may send the EID or the Called Party Number (CdPN) in the SETUP message to the terminal for the purpose of terminal selection.

SPIDs should not be confused with TEIs (terminal endpoint identifiers). TEIs identify the terminal at Layer 2 for a particular interface (line). TEIs will be unique on an interface, whereas SPIDs will be unique on the whole switch and tend to be derived from the primary directory number of the subscriber. Although they are used at different layers, they have a one-to-one correspondence, so mixing them up isn't too dangerous. TEIs are dynamic (different each time the terminal is plugged into the switch), but SPIDS are not. Following the initialization sequence mentioned above, the one-to-one correspondence is established. TEIs are usually not visible to the ISDN user, so they are not as well-known as SPIDs.

The "address" of the layer 3 message is usually considered to be the Call Reference Value (also dynamic, but this time on a per-call basis) as opposed to the SPID, so the management entity in the ISDN device's software must associate EID/CdPN on a particular TEI and Call Reference Number to a SPID.

There are some standards that call for a default Service Profile, where a terminal doesn't need to provide a SPID to become active. Without the SPID, however, the switch has no way of knowing which terminal is which on the interface, so for multiple terminals an incoming call would be offered to the first terminal that responded, rather than to a specific terminal.

Can I purchase European ISDN devices and use them successfully in the U.S.?

There are four major problem areas regarding interoperability of ISDN equipment between countries.

The first has to do with voice encoding, and is only a problem if the equipment is a telephone. Equipment designed for use in North America and Japan uses mu-law encoding when converting from analog to digital, whereas the rest of the world uses A-law. If the equipment has a switch for selecting one or the other of these encoding types, then there will not be a problem with the voice encoding.

The second has to do with the way the equipment communicates with the telephone exchange. There are interoperability problems because there are so many different services (and related parameters) that the user can request, and each country can decide whether or not to allow the telephone exchange to offer a given service. Also, the specifications that describe the services are open to interpretation in many different ways. So, as with other interoperability problems, you must work with the vendors to determine if the equipment will interoperate. This is a basic problem; it affects all ISDN equipment, not just voice equipment.

The third has to do with homologation, or regulatory approval. In most countries in the world the manufacturer of telephone equipment must obtain approvals before the equipment may be connected to the network. So, even if the equipment works with the network in a particular country, it isn't OK to hook it up until the manufacturer has jumped through the various hoops to demonstrate safety and compliance. It is typically more expensive to obtain world-wide homologation approvals for a newly developed piece of ISDN equipment than it is to develop it and tool up to manufacture it.

A fourth issue is that in the U.S., the TA and NT1 are both provided by the customer, while in Europe the NT1 is provided by telephone company. Stated differently, if you walk into a store in the U.S. and buy something to plug into an ISDN line, it may be designed as a one-piece unit that connects to point U. In Europe, you would get something that plugs into point T. Thus, you might take a piece of U.S.-originated equipment to Europe and find that it won't work because the jack in Europe is a T interface and the plug on your U.S. equipment is a U interface.

There are attempts to remedy this situation, particularly for BRI ISDN. In North America, the National ISDN User's Forum is coming up with standards to increase the uniformity of ISDN services. In Europe, a new standard called NET3 is being developed.

Do different manufacturers' terminal adapters interoperate when used asynchronously?

There is a standard up to 19.2K (V.110), but above that there is no real standard implemented. However, in practice there is a fair degree of interoperability (even when the TA's manual tells you otherwise) because many TAs use the same chip set (supplied by Siemens) which happily goes up to 38.4. TAs from different suppliers that are using the Siemens chips have a fair chance of interoperating at up to 38.4k.

What are the basic differences between a bonded 128KB ISDN connection and a switched 56 line?

Switched 56 and ISDN are both dial-up services. Given a choice of ISDN or Sw56, go with ISDN. Switched 56 is an earlier, predecessor service which is being phased out in favor of ISDN. The two interwork freely (call each other), but ISDN gives you two B channels for usually a lower price than one Sw56 line. The delay is about the same, unless you're closer to one service's switch than the other.

How long should call setup take when using a TA?

The "less than a second" call setup sometimes claimed seems to be rare. TAs have a negotiation phase and it typically takes around four seconds to get through to the remote site.

Appendix A: ISDN Dial-Tone Service Providers

Bellcore (Bell Communications Research)

National ISDN HotLine:	1-800-992-ISDN
Fax:	201.829.2263
E-Mail:	isdn@cc.bellcore.com
URL:	http://info.bellcore.com
System prompt:	ftp info.bellcore.com

Company	Contact	Telephone no.
Ameritech	National ISDN Hotline	1-800-TEAMDATA (1-800-832-6328)
Bell Atlantic In N.J., call your local telephone office.	SDN Sales & Tech Ctrl For Small Businesses	1-800-570-ISDN (1-800-570-4736) 1-800-843-2255
Bellsouth	ISDN HotLine	1-800-428-ISDN (1-800-428-4736)
Cincinnati Bell	ISDN Service Center	513-566-DATA (513-566-3282)

GTE	Menu-driven information	1-800-4GTE-SW5
	Florida, North Carolina, Virginia, and Kentucky:	1-800-483-5200
	Illinois, Indiana, Ohio, and Penn.	1-800-483-5600
	Oregon and Washington	1-800-483-5100
	California	1-800-483-5000
	Hawaii	1-800-643-4411
	Texas	1-800-483-5400
Nevada Bell	Small business	702-333-4811
	Large business	702-688-7100
	ISDN Sales Hotline	1-800-GET-ISDN
Nynex		1-800-438-4736
	New England States	617-743-2466
PACIFIC BELL	ISDN Service Center	1-800-4PB-ISDN
		1-800-472-4736
	24 Hr. Automated ISDN Avail. Hotline	1-800-995-0346
	ISDN Telemarketing	1-800-662-0735
Rochester Telephone	ISDN Information	716-777-1234
SNET	Donovan Dillon	203-553-2369
Stentor (Canada)	ISDN "Facts by Fax"	1-800-578-ISDN
	Steve Finlay	604-654-7504
	Glen Duxbury	403-945-8130
Southwestern Bell	Austin, TX	1-800-SWB-ISDN
	Dallas, TX	214-268-1403
	North Houston, TX	713-537-3930
	South Houston, TX	713-567-4300
	San Antonio, TX	210-351-8050
	For ISDN availability for remaining locations pls. contact the Bellcore ISDN HotLine at	1-800-992-ISDN
U. S. West	Ron Miller	303-965-7153
	Ron Woldeit	206-447-4029
	Denver, CO	1-800-246-5226
	Julia Evans	303-896-8370

National ISDN Long-Distance Carrier Contacts

AT&T	AT&T Front End Center	1-800-222-7956
GTE	Nationwide availability/pricing	1-800-888-8799
	Ron Sterreneberg	214-718-5608
MCI	Tony Hylton	214-701-6745
	ISDN Availability	1-800-MCI-ISDN
U.S. Sprint	Rick Simonson	913-624-4162
Wiltel	Justin Remington	918-588-5069

Appendix B

Vendor Information

AccessWorks Communications Inc
670 North Beers Street
Holmdel, NJ 07733
Tel: 800 248 8204 or 908 721 1337
Fax: 908 888 4456
Internet: info@accessworks.com

Adtran, Inc.
901 Explorer Blvd
Huntsville, AL 35806-2807 USA
Tel: +1 205 971 8000
Fax +1 205 971 8030

Advanced Micro Devices
901 Thomson place
Mailstop 126
Sunnyvale, CA 94086
Tel: (408) 732 2400

ANDO:
7617 Standish Place
Rockville, MD 20855
Tel: (301) 294-3365

Fax: (301) 294-3359
Internet: mgriffin@access.digex.net

Ascend Communications, Inc.
1275 Harbor Bay Pkwy
Alameda, CA 94501
Tel: (510) 769-6001
Internet: info@ascend.com

AT&T
1-800-222-PART: Quick access to small quantity orders of ISDN products.
Personal Desktop Video or TeleMedia Connection System:
Visual Communications Products
8100 East Maplewood Avenue 1st Floor
Englewood, CO 80111
Tel: (800) 843-3646 or (800)VIDEO-GO Prompt 3

AT&T Microelectronics
Allentown, PA
Tel: (800) 372-2447
Distributor: CoSystems at 408.748.2190
Marketing: Steve Martinez at 408.748.2194 (steve@cosystems.com)
Technical issues: Gary Martin at 408.748.2195 (gary@cosystems.com)

BinTec Computersysteme GmbH
Willstaetter Str. 30
D-90449 Nuernberg Germany
Tel: +49.911.9673-0
Fax: +49.911.6880725
Internet: vertrieb@bintec.de

Cisco Systems
170 West Tasman Drive
San Jose, CA 95134
Tel: (800) 553-6387 or (408) 526-4000
Fax: (408) 526-6387

Combinet
333 West El Camino Real, Suite 240

Sunnyvale, California 94087
Tel: (408) 522 9020 or (800) 967-6651 for product lit
Fax: (408) 732 5479 (fax)

Conware Computer Consulting GmbH
Killisfeldstr. 64
D-76227 Karlsruhe Germany
Phone: +49.721.9495-0
Fax: +49.721.9495-130
Internet: vertrieb@conware.de

CPV-Stollmann Vertriebs GmbH
Gasstrasse 18
D-22761 Hamburg Germany
P.O. Box 50 14 03
D-22714 Hamburg Germany
Tel: +49-40-890 88-0
Fax: +49-40-890 88-444
Internet:
Info@Stollmann.DE (general inquiries)
Helge.Oldach@Stollmann.DE (IPX router technical contact)
Michael.Gruen@Stollmann.DE (IP router technical contact)

CSI (Connective Strategies, Inc.)
Clyde Heintzelman, V. P. Marketing
4500 Southgate Pl., Suite 100
Chantilly, VA 22021
Tel: (703) 802-0023
Fax: (703) 802-0026
Internet: info@csisdn.com

diehl isdn GmbH
Bahnhofstrasse 63
D-7250 Leonberg Germany
Tel. +49-7152-93 29-0
Fax. +49-7152-93 29-99
Internet: bode@diehl.de

DigiBoard
6400 Flying Cloud Drive

Eden Prarie, MN 55344
Tel: (612) 943 9020 or (800)-344-4273
Fax: (612) 643 5398
Internet: info@digibd.com

Digital Equipment Co
REO2 G/H2 DEC Park
Worton Grange
Reading, Berkshire, England

DGM&S
609.866.1212

EICON Technology Sales Office
14755 Preston Road, Suite 620
Dallas, TX 75240
Tel: (214) 239-3200
Fax: (214) 239-3304
EiconCard ISDN/PC ISA PC card. Supports 2B + D "multiplexed" over a single RJ45 connector. They provide s/w for Windows, OS/2, SCO UNIX, UNIX SVR4, Netware. List price $1395.

EuRoNIS
166 rue Montmartre
75002 Paris, France
Tel: +33 (1) 44 82 70 00
Fax: + 33(1) 42 33 40 98
Internet: euronis@applelink.apple.com
Manufacturer of the Macintosh Planet-ISDN NuBus Card.

Gandalf Technologies
130 Colonnade Road South
Nepean, Ontario, Canada K2E 7M4
Tel: (800) GANDALF or (613) 723-6500
Fax: (613) 228-9510

Hayes ISDN Technologies
501 Second St., Suite 300
San Francisco, CA 94107

Tel: (415) 974-5544
Fax: (415) 543-5810

Hermstedt GmbH
Kaefertaler Strasse 164
D-68167 Mannheim Germany
Phone: +49 (621) 3 38 16-0
Fax: +49 (621) 3 38 16-12

International Business Machines
Networking Systems Division
3039 Cornwallis Road
Research Triangle Park, NC 27709
Tel: (800) 426-2255 or (919) 543-7421
Fax: (919) 543-5417

INS (Inter Networking Systems)
P.O. Box 101312
D-44543 Castrop-Rauxel Germany
Tel: +49 2305 356505
Fax: +49 2305 24511
Internet: info@ins.de

Intel Corporation
Intel Products Group
5200 N.E. Elam Young Parkway
Hillsboro, Oregon, 97124-6497
FaxBack 1-800-525-3019
Product Info:
(800)-538-3373, in the US and Canada
+44-1793-431155, in Europe
+1-(503) 264-7354, worldwide
Intel BBS (503) 264-7999 (modem settings 8-N-1, up to 14.4Kbps)
Tech. support (503) 629-7000

ISDN Systems Corporation
8320 Old Courthouse Road, Suite 200
Vienna, VA 22182
Tel: (703)-883-0933

MERGE Technologies Group, Inc.
211 Gateway Road West, Ste. 201
Napa, CA 94558
Tel (800) 824-7763
Fax: (707)252-6687

MITEL Corporation
360 Legget Drive
Kanata, Ontario, Canada K2K 1X3
Tel: (613) 592-2122 (ask for Paul Mannone or Peter Merriman)

Motorola UDS
5000 Bradford Drive
Huntsville, AL 35805
Tel: (205) 430 8000

Ms Telematica
via S. Marcellina 8
20125 Milano Italy
Tel: +39.2.66102315
Fax: +39.2.66102708
Internet: mstelema@icil64.cilea.it

netCS Informationstechnik GmbH
Feuerbachstr. 47-49
12163 Berlin 41 Germany
Tel: +49.30/856 999-0
Fax: +49.30/855 52 18
Internet: sales@netcs.com or support@netcs.com

Network Express, Incorporated
World Headquarters
4251 Plymouth Road
Ann Arbor, MI 48105
Tel: (313) 761-5005
Fax: (313) 995-1114
Internet: info@nei.com

Paxdata Networks Limited
Communications House

Frogmore Road
Hemel Hempstead HERTS HP3 9RW UK
Tel: 0442 236336
Fax: 0442 236343
Marketing: Jim Fitzpatrick (jim@paxdata.demon.co.uk)
Technical: Giles Heron (giles@paxdata.demon.co.uk)

SO ISDN NuBus card
Siemens Components Inc.
Integrated Circuit Division
2191 Laurelwood Road
Santa Clara, CA 95054-1514
Tel: (408) 980-4500

Spider Systems
Spider Systems Limited
Spider House
Peach Street
Wokingham, England RG11 1XH
Tel: 0734 771055
Fax: 0734 771214

Sun Microsystems Computer Company (SMCC)
2550 Garcia Avenue
Mountain View, CA 94043
Tel: (800) USA-4SUN

Telenetworks
Lauren May/Bob Gefvert
625 Second St., Suite 100
Petaluma CA 94952
Tel: (707) 778-8737
Fax: (707) 778-7476
Internet: info@tn.com

Teleos
2 Meridian Road
Eatontown, NJ 07724
908.389.5700

Telesoft International, Inc.
4029 South Capital of Texas Highway
Austin, TX 78704
Tel: (512) 373-4224

Telrad Telecommunications, Inc.
135 Crossways Park Drive
Woodbury, New York 11797
Tel: (516) 921-8300 or (800) 645-1350
TelradPAC: 0B+D PAD, NI-1 & Euro-ISDN; IDS: V&D phone, NI-1; MTA: V.110 TA, Euro-ISDN

TPI (Tele-Path Industries, Inc.)
221 South Yorkshire Street
Salem, VA 24153
Tel: (703) 375 0500

Trillium
Tel: (310) 479-0500
Internet: marketing@trillium.com

Zydacron, Inc.
670 Commercial Street
Manchester, NH 03101
Tel: (603) 647-1000
Fax: (603) 647-9470

Appendix C

ISDN Bibliography and On-line Resources

Our goal in providing this appendix is not to tell you how to use an on-line information service or the Internet in general, or CompuServe or the World-Wide Web in particular. Nor do we want to provide the be-all and end-all of ISDN resources. Rather, we just want to tell you what information is available on CompuServe and the Internet, what it's made up of, and why you might find it interesting. We also want to point you toward those books and materials about ISDN that we've found most useful in the course of researching and writing this book.

This appendix focuses first on what's up on CompuServe and the Internet, how best to interact with it, and what kinds of things you can and cannot find up there. It tells you how to be effective when you work with CompuServe or the Internet, from the standpoint of knowing what to look for, which kinds of questions you can ask, and the answers you're likely to get. It also helps you to understand just what kind of help you can expect to get from the on-line community and what to do if you can't get the help you need. Then, at the conclusion of this appendix, we include an ISDN section at the head of the "Bibliographies and Resources" section.

By now, you've probably noticed that we've mentioned only CompuServe and the Internet as sources of on-line information. "What about the others?" you might ask. Yes, we know there's also America On-Line, Prodigy, GEnie, and a bunch of other lesser contenders in this field. But none of them have staked a presence in the area of technical information and support on-line as CompuServe has, nor does any of them have the breadth and reach of the Internet (which can't yet compare with CompuServe's depth of offerings, but is quickly catching up).

That's why we focus the bulk of our discussion in this appendix on these two information sources, even though there are more to choose from. In the next-to-last section, entitled "Other On-line Resources," we'll try to give you some ideas about other places worth looking, but this will be a set of cursory suggestions, rather than an in-depth investigation.

What Does "On-line" Really Mean?

In the context of our discussion, "online" means that you have to log in to somebody else's network (frequently using a modem) to access their information collection rather than your own network. While this may sound inconvenient—and it sometimes is—the benefits invariably outweigh any inconvenience, costs, and effort that might be involved.

For the record, these benefits include the following:

- Free access to technical support operations for questions and answers via forums (CompuServe) or newsgroups and mailing lists (the Internet). Even if you never ask a question yourself, reading other people's questions (and the answers that go with them) can be enormously informative.
- Access to on-line sources for software patches and fixes for a broad range of products. Rather than waiting for the vendor to send you a disk, or paying long-distance charges to access their private bulletin board, you can get the latest versions of software (or the tools to turn your software into the latest version) with a local phone call and a (sometimes lengthy) download.
- Access to shareware and freeware that can extend your network's capabilities, or increase your personal productivity. Much of the software this book's authors use for things such as screen shots, graphics, file compression, and more originated on the Internet or Compuserve. A little prospecting can work wonders in this area!
- The biggest benefit by far is the opportunity to meet and interact with your peers and colleagues in the networking profession, and to learn from other people's experiences and mistakes. You'll also have the occasional chance to learn from the wisdom of real experts, in-

cluding the developers of the software or hardware you're using, or world-renowned gurus from a variety of fields.

All in all, there's a lot to be gained from going on-line to look for information on just about any subject, but especially for technical and computer-related subjects. Since that's where networking fits pretty neatly, these resources are excellent (some would argue, indispensable) sources for information on the whole gamut of networking topics, products, technologies, and issues.

Now that we've gotten you all excited about the possibilities inherent in online information access, let's talk about the costs. Whether you join up with CompuServe, get onto the Internet, or—like this book's authors—do both; you can't join up without incurring some costs.

For CompuServe, this involves a series of account options, with associated monthly fees and additional charges for on-line time (usually above a certain number of "free" monthly hours). A light user shouldn't have to spend more than $10–15 a month for the service, but if you make regular downloads or spend significant amounts of time online, it's easy to spend $50–100 a month, or more.

For the Internet, you'll have to arrange for a connection with an Internet Service Provider (ISP), and select one of the many options available for an Internet connection. For individuals or small businesses, we recommend using ISDN—or at least a V.34 modem—with a PPP connection. While prices vary from location to location, you should expect to pay between $80 and $150 a month for ISDN service with this kind of a connection, or about half that with a dial-up POTS connection. This usually entitles you to 10–20 hours per month of "free" online time, after which an hourly fee will be charged for additional hours.

There are lots of other Internet account options available from most ISPs, which can vary from dial on demand to dedicated accounts, or according to the bandwidth of the connection involved (modem, ISDN, T1, T3, etc.). If you're interested in attaching your network to the Internet, or need more bandwidth than an ISDN connection can provide, talk to your local ISPs, or to national ISPs that offer service in your area. If you shop carefully for the best combination of price and service, you should be able to find something you can live with!

As with any other service, whether it's CompuServe, the Internet, or both, you'll want to do your best to learn how to use these information conduits effectively, to get the best bang for your bucks. Please consult the bibliographies at the end of this Appendix for a list of resources that can help you learn what it takes to get the best use of either or both of these services.

CIS: THE COMPUSERVE INFORMATION SERVICE

The CompuServe Information Service (CIS) is an electronic information service that offers a selection of thousands of topics for your perusal.

CompuServe, a for-a-fee service, requires an individual account (called a membership number) with an accompanying password to be accessed. There are many ways to obtain trial access at no charge, but if you want to play on CompuServe, sooner or later you have to pay for the privilege. CompuServe charges a monthly membership fee, in addition to a fee for connection time. Some of the services available on CompuServe have additional charges as well. Be warned! It's easy to spend time ó and money ó on CompuServe.

Forums for Conversation and Investigation

When you access CompuServe, it's necessary to select an area of interest to focus your exploration of the information treasures available. On CompuServe, information is organized into forums. A forum is an area dedicated to a particular subject or a collection of related subjects, and each forum contains one or more of the following:

- **Message board**: Features electronic conversations organized by specific subjects into sections related to particular topics (ISDN or the name of an ISDN vendor is probably what you'd look for as the focus for a section or forum on ISDN). A given sequence of messages, chained together by a common subject or by replies to an original message, is called a thread. It's important to notice that threads may read like conversations but that messages in a thread can be separated from one another by hours or days. Following threads is a favorite pastime for those who spend time on CompuServe.
- **Conference room**: An electronic analogue to the real thing, it brings individuals together to exchange ideas and information in real time. It's much like a conference telephone call except that, rather than talk to each other, the participants communicate by typing on their

keyboards. Conference rooms are not for the faint of heart, and they can be frustrating for those with limited touch-typing skills.

- **File library**: A collection of files organized by subject that can be copied ("downloaded" is the CompuServe term) for further perusal and use. Examples of file types found in CompuServe libraries include archived collections of interesting threads, documents of all kinds, and a variety of software ranging from patches and fixes for programs to entire programs.

In all, many, many worlds of information are available on CompuServe, any or all of which can by themselves be a completely absorbing source of information, gossip, software, and activity. With all its elements taken together, CompuServe is a perfect example of what might be called an "electronic information warehouse."

Getting a CompuServe Membership

You can obtain an account over the telephone or by writing to CompuServe and requesting a membership. For telephone inquiries, ask for Representative 200. Here are the numbers to use:

- Within the U.S. (except Ohio), including Alaska, Hawaii, Puerto Rico, and the American Virgin Islands, call toll-free at 800-848-8199.
- Outside the U.S., in Canada, and in Ohio, call 614-457-8650.

Telephone hours are from 8 A.M. to 10 P.M. Eastern time Monday through Friday, and from noon to 5 P.M. on Saturday. Written inquiries for a CompuServe account should be directed to:

CompuServe, Inc.
Attn: Customer Service
P.O. Box 20212
5000 Arlington Centre Boulevard
Columbus, OH 43220
U.S.A.

Accessing CompuServe

To get access to CompuServe, you must equip your computer with a modem and attach that modem to a telephone line. You also need some kind of communications program, to let your computer "talk" to CompuServe by using the modem and to help you find your way around its on-line universe. Finally, you have to obtain a telephone number for CompuServe—most of them are local numbers, especially in the U.S.— that's appropriate for the type and speed of modem you're using. At this point, CompuServe isn't available via ISDN outside the Columbus, Ohio, area (CompuServe's home town), but keep your eyes peeled, because that kind of access is surely just a matter of time.

At present, though, connection-time charges are based on how fast your modem is—faster modems cost more. However, the higher charges are typically offset by even faster transfer speeds. If your CompuServe bill is $30 a month or higher, most high-speed modems will pay for themselves in six months or less based on the reductions in fees you realize by using one.

After you are connected to CompuServe, you enter your membership number and your password. First-time users should follow the instructions provided by your CompuServe representative or in the CompuServe Starter Kit that's available from CompuServe (for an additional fee).

After you're logged in, getting directly to a named CompuServe forum is easy-as long as you know the name of the forum you're after. When you simply type **GO <NAME>** from the CompuServe prompt, you are presented with a menu of additional choices for that forum. To get started with your information mining efforts, use the **FIND** command with ISDN, a vendor's name, or a product name, whatever you're after. This will normally produce a list of forums that you can visit to further explore potential sources of information.

The Many Forums of CompuServe

There are plenty of vendors and networking communities represented on CompuServe. You'll find a rich selection of Novell forums (**GO NetWire**), and Microsoft-related forums and libraries, for everything from vanilla Windows, to Windows for Workgroups, Windows95, and Windows NT Advanced Server (use **GO MICROSOFT** to get to the root of the Microsoft forums; don't skip the Microsoft Knowledge Base at **GO MSKB**, either).

You'll also find plenty of IBM-related forums (**GO IBM**) or OS/2-related information (**GO OS/2**). Don't forget to use the FIND command to locate other vendors and products, as well.

The Internet

The Internet, as a wag might put it, is a "whole ënother story." There are more riches to be found on the Internet than you could shake a stick at. This won't stop us from pointing you at a few good stops along the way, but it will effectively prevent us from covering all the possible bases. Rather than trying to tell you where all the goodies are, we're going to explain how to search for the information you seek.

Our primary approach to the Internet requires that you have access to the World Wide Web (WWW), usually known as "the Web." The Web is a worldwide collection of hypertext information servers that is made easy to navigate through the use of hypertext links, that let you jump effortlessly from document to document (or within a document) simply by activating a link on a document you're examining (and for most users, this requires no more effort than double-clicking a word or graphic on your display). Secondarily, access to electronic mail and/or USENET newsgroups will be quite helpful as well.

Searching for Satisfaction

In much the same way that the FIND command on CompuServe lets you ask for information by company or product name, the Web sports a number of database front-ends called "search engines," that will let you enter a keyword (or several, in fact) for search. These programs will return a collection of hypertext links to sites that match your keywords to various locations on the Web, ready for you to double-click on them and investigate further.

The name used to attach to a Web resources is called a "URL" (an acronym for Uniform Resource Locator, a way of designating sites and information accessible through the Web). Here are the URLs for a handful of popular—and useful—Web search engines. If you simply point your Web browser at one of these, you'll then be able to get pointers to the information you're looking for.

Sometimes, using the right tools can make using the World Wide Web for research much simpler. There is a class of software tools called *search engines* that can examine huge amounts of information to help you locate Web sites of potential interest. Here's how most of them work:

- Somewhere in the background, laboring in patient anonymity, you'll find automated Web-traversing programs, often called *robots* or *spiders*, that do nothing but follow link after link around the Web ad infinitum. Each time they get to a new Web document, they peruse and catalog its contents, storing the information up for transmission to a database elsewhere on the Web.
- At regular intervals, these automated information gatherers will transmit their recent acquisitions to their parent database, where the information is sifted, categorized, and stored.
- When you run a search engine, you're actually searching the database that's been compiled and managed through the initial efforts of the robots and spiders, but which is handled by a fully functional database management system that communicates with a customized program for your search form.
- Using the keywords or search terms you provide to the form, the database locates "hits" (exact matches) and also "near-hits" (matches with less than the full set of terms supplied, or based on educated guesses about what you're *really* trying to locate).
- The hits are returned to the background search program by the database, where they are transformed into a Web document to return the results of the search for your perusal.

If you're lucky, all this activity will produce references to some materials that you can actually use!

The Search Engines of (our) Choice

We'd like to share some pointers to our favorite search engines with you, which you'll find in Table 1. This is not an exhaustive catalog of such tools, but all of them will produce interesting results if you use "CGI" or "CGI scripts" as search input.

Search engine name and info	URL
EINet Galaxy MCI spinoff EINet's engine	http://www.einet.netLycos
Carnegie-Mellon engine W3 Org Virtual Library	http://lycos.cs.cmu.edu
W3 Org outsourced project Wandex	http://www.stars.com
MIT spinoff's engine WebCrawler	http://www.netgen.com/cgi/wandexWebCrawler
University of Washington engine World Wide Web Worm (WWWW)	http://webcrawler.cs.washington.edu/ WebCrawler/WebQuery.html
University of Colorado engine	http://www.cs.colorado.edu:80/home/mcbryan/ WWWW.htm
Yahool	http://www.yahoo.com

Table C.1 *Web search engines.*

When you're using these search tools, the most important thing to remember is that the more specific you can make your search request, the more directly related the results will be to what your looking for. Thus, if you're looking for information about ISDN terminal adapters, you might try using "ISDN TA" or "TA" as your search terms instead of simply using "ISDN." While you may get plenty of nothing when using search terms that are too specific, that's better than looking through a plenitude of irrelevant materials when nothing is all that's in there!

The Web has to be experienced to be believed. Since the authors' initial exposure to it about two years ago, it's completely changed the way we approach research of any kind. We hope you'll find it to be useful, but we must warn you—it's also completely addicting!

Other Ways to Get Internet Satisfaction

When it comes to classifying the kinds of information you'll encounter on the Internet in any search for networking information, specifications, and examples, here's what you're most likely to find:

Focused Newsgroups

Focused newsgroups are basically congregations of interested individuals, who congregate around a specific topic on USENET, BITNET, or one of the other regular message exchange areas on the Internet.

Where networking is concerned, this involves a handful of primarily USENET newsgroups with varying levels of interest in (and coverage of) networking- or vendor-specific or related topics, such as NetWare, LAN Server, or NTAS, networking protocols or products, related technologies such as Ethernet, Token-Ring, or FDDI, and other related areas.

To begin with, you'll want to obtain a list of the newsgroups that your Internet Service Provider (ISP) carries. Normally, you will already have access to this list through whatever newsreader you're using, but you can usually get a plain-text version of this list just by asking for it.

Then, take this plain-text file and open it with your favorite editor or word processor that contains a search command. By entering the name of the company, technology, or product that you're interested in, you can see if there are any newsgroups devoted to its coverage (a recent check on our part discovered several hits for the term "isdn" including *comp.dcom.isdn*, *de.comm.isdn*, *fido.ger.isdn*, and *relog.isdn*). Of these, two are aimed at German audiences (de.comm.isdn, and fido.ger.idsn), we use *comp.dcom.isdn* as a terrific source of information, and know nothing about the *relog.isdn* list, since we found it empty.

In fact, the only way to tell if a newsgroup can do you any good is to drop in for a while and read its traffic. You should be able to tell, in a day or two, if the topics and coverage are interesting and informative. If they are, you should consider subscribing to the newsgroup, or at least dropping in from time to time to read the traffic. Remember, too, that these newsgroups are a great source of technical information, and that they often have vendor technical support employees assigned to read them, ready to answer technical questions on your behalf.

Focused Mailing Lists

Focused mailing lists originate from targeted mail servers that collect message traffic from active correspondents, and then broadcast the accumulated traffic to anyone who signs up for the mailing list.

Entering and leaving a mailing list takes a little more effort than subscribing to or leaving a USENET newsgroup, but otherwise, these two categories provide the same kind of information: daily message traffic—sometimes quite voluminous—focused on networking or related topics.

Locating mailing lists can sometimes be tricky. While you often learn about them only by reading message traffic on newsgroups, you can sometimes find them mentioned in search engine output, or by asking a users' group or a technical support person focused on a particular topic or area. Even so, they can be incredibly useful.

Information Collections from "Interested Parties"

Sometimes individuals with special interests in a particular area—such as networking—will collect information about their area of concern, and publish it in a variety of forms that can range from Web pages to file archives available on private or public servers.

While such collections can often be eclectic and idiosyncratic, the best of them can offer outstanding "jumping-off points" for investigating any particular topic. This is as true for networking as it is for other topics.

As with mailing lists, finding these gems can be a matter of hit or miss. By watching the message traffic on newsgroups or mailing lists, you'll figure out who the gurus or forward-looking individuals are. By looking in their messages for pointers to Web pages or other resources (which you'll often find in the .sig, or signature files, at the end of their messages), you can sometimes get pointers to great sources of information.

In the same vein, if you see that a particular individual is a consistent and reliable source of good information on a particular topic, send him or her an e-mail message and request a list of recommended on-line resources. While you may not always get a response (some of these people are very busy), it never hurts to ask, and the occasional answer can provide a real treasure trove of information pointers!

The best source of ISDN information on the Internet, bar none, may be found at the following WWW URL:

http://www.alumni.caltech.edu/~dank/isdn/

This is Dan Kegel's ISDN Web page, and it's got pointers to everything worth knowing about ISDN on the Internet (no kidding!). If you can get here, you can get all the online information about ISDN that you'll ever need.

Information from Special Interest Groups

Special interest groups cover a multitude of approaches to their topics: They can be trade or industry organizations, research or standards groups, or even companies involved in particular activities.

Often, the groups with vested interests in a technology will provide information on that technology, along with pointers to other sources as well. This is as true for networking as it is for other topics, but because these groups are nonpareils of Web and Internet presence, they are often among the best places to start looking.

It's often been said that "It's not *what* you know, it's *who* you know, that counts." When it comes to locating Internet resources, this may sometimes seem more like "*where* you know," but the principle remains pretty much the same. Thus, for particular topics, you shouldn't point your search engine only at company, product, or technology names; try pointing them at the names of such groups as well. Here again, these can be incredible sources of useful information.

Dan Kegel's page includes a section entitled "ISDN User Groups;" look here for information about local and national groups, most of which have their own mailing lists, file archives, and Web servers that you can explore for good ISDN information (and contacts for consultants, equipment dealers, etc.).

OTHER ONLINE RESOURCES

Many companies that are too small, too poor, or otherwise disinclined to participate on CompuServe or the Internet will maintain their own private Bulletin Board Systems (BBSs), which you can dial up and investigate. These are always free, but they're also almost always long-distance calls, so what you save in connect time costs to a service provider you'll probably end up paying to your long-distance company instead.

Nevertheless, when other avenues fail to turn up what you're looking for, it's a good idea to call the vendor and ask if they offer a BBS. This kind of setup often gives the opportunity to communicate with technical support via e-mail, instead of enduring "eternal hold" while waiting to speak to a real human being, and can often provide direct access to software patches, fixes, upgrades, FAQs, and other kinds of useful documentation.

In the same vein, many companies offer faxback services that can ship paper-based documentation, order forms, and other goodies to your fax-machine. This has the advantage of costing you only as much long-distance time as it takes you to request the information you're after; after that, the transmission costs are usually borne by the vendor.

Digging for information is an endeavor where persistence usually pays off. If you're bound and determined to ge the facts, figures, or help you need, you'll eventually be able to get it. Just be sure to leave no resource uninvestigated, no possible avenue untraveled, and no stone unturned!

Non-computerized Resources Worth Investigating

Even though they may not be as dynamic and interactive as on-line resources, don't overlook the information you can glean from more conventional paper-based publications. (We know you've got to be somewhat open-minded in this regard, because you're reading this book!) Nevertheless, we'll do our best to acquaint you with some books, magazines, and publishers to check out in your quest for the latest and greatest networking information. You should find this information in the bibliographies at the end of this appendix, and on Dan Kegel's page (check the listings under the title "ISDN Periodicals and Magazines" for some useful resources).

SUMMARY

In this appendix, we've tried to point you at the best and brightest of on-line (and other) information resources. Over the years, we've learned that both CompuServe and the Internet are essential to our research, but you will probably be able to get by with only one or the other. Whatever your choice of on-line information, however, we're sure you'll become dependent on it in no time (if you aren't already). It's definitely one of those things that, as soon as it becomes familiar, you wonder how you ever managed without it! In the bibliographies that follow, we try to provide some tools to bring you up to speed quickly enough to make the investment pay off right away.

Bibliographies and Resources

ISDN

Hopkins, Gerald L.: *The ISDN Literacy Book*: Copyright 1994, Addison-Wesley Publishing Company, Inc., Reading, MA. A great comprehensive introduction to ISDN for managers and technicians alike.

Kessler, Gary C.: *ISDN Second Edition*: Copyright 1993, McGraw-Hill, Inc., New York, NY. Lots of practical information on ISDN. This book could serve as a complete guide.

Motorola University Press: *The Basics Book of ISDN*: Copyright 1992, Motorola, Inc., published by Addison-Wesley Publishing Company, Inc., Reading, MA. A short and useful introduction to ISDN terms and technology.

Stallings, William: *ISDN, An Introduction*: Copyright 1989, Macmillan Publishing Company, a division of Macmillan, Inc., New York, NY. Stallings has a prodigious understanding of ISDN. An exhaustive examination of ISDN from the high levels to the nitty gritty technical details.

General

Bernard Aboba: *The Online User's Encyclopedia*: Copyright, 1993, Addison-Wesley Publishing Co, Reading, MA. A general book that covers online topics from A to Z, this tome defies description but is incredibly useful.

CompuServe

Chappell, Laura: *Using Novell's NetWire*: Copyright, 1992, Know, Inc., Provo, UT. A dated but useful overview of NetWire's composition and uses.

Tidrow, Rob, Jim Ness, Bob Retelle, and Chen Robinson: *New Riders' Official CompuServe Yellow Pages*: Copyright, 1994, New Riders Publishing, Indianapolis, IN. A directory to CompuServe resources organized like the Bell Yellow Pages, with great sections on vendors and networking.)Warning: be sure to get a current edition, as this information goes stale quickly.)

Wagner, Richard: *Inside CompuServe, 2nd Edition*: Copyright 1994, New Riders Publishing, Indianapolis, IN. A useful overview of how CompuServe behaves, what kinds of access software are worth considering, and what sorts of resources it contains.

Wiggins, Robert and Ed Tittel: *The Trail Guide to CompuServe*: Copyright, 1994, Addison-Wesley Publishing Company, Reading, MA. A quick overview of the CompuServe Information Manager (CIM) software for Windows and Macintosh, and a quick, but useful, guide to resources online. Includes a chapter on network-related forums and topics.

Wang, Wallace: *CompuServe for Dummies*: Copyright, 1994, IDG Books Worldwide, Indianapolis, IN. One of the best all-round resources on CompuServe available, this book covers software, organization, and effective "surfing" techniques.

Internet

Angell, David, and Brent Heslop: *Mosaic for Dummies*, Windows Edition: Copyright, 1995, IDG Books Worldwide, Indianapolis, IN. Excellent coverage of the Mosaic Web browser, and Web-based online resources.

December, John, and Neil Randall: *The World Wide Web Unleashed*: Copyright, 1994, SAMS Publishing, Indianapolis, IN. The best of the general WWW reference books, this one covers all the topics, including one of the most comprehensive guides to online resources we've ever seen anywhere.

Dern, Daniel: *The Internet Guide for New Users*: Copyright, 1994, McGraw-Hill, Inc., New York, NY. One of the three best all-around Internet books, this one covers a little bit of everything, including programs to use and places to look for information.

Hahn, Harley, and Rick Stout: *The Internet Complete Reference*, Osborne/McGraw-Hill, Inc., Berkeley, CA. Another of the three best all-around Internet books, this one is aimed more at intermediate to advanced users, but covers a lot of ground anyway.

Krol, Ed: *The Whole Internet User's Guide*, 2nd Edition, O'Reilly & Associates, Inc., Sebastopol, CA. The third of the three best Internet books around, this was the earliest and is still a personal favorite.

Levine, John R. and Carol Baroudi: *The Internet for Dummies, 2nd Edition*: Copyright, 1994, IDG Books Worldwide, Indianapolis, IN. An excellent overview of the Internet's many protocols, programs, and capabilities.

Levine, John R., and Margaret Levine Young: *More Internet for Dummies*: Copyright, 1994, IDG Books Worldwide, Indianapolis, IN. A continuation of the coverage in the first book, this volume provides a good introduction to the World Wide Web and how to use it.

Tittel, Ed and Margaret Robbins: *Internet Access Essentials*: Copyright, 1994, Academic Press Professional, Boston, MA. The fourth of the three best all-around Internet books, this one was co-written by the author, which means he thinks it's pretty darned good indeed!

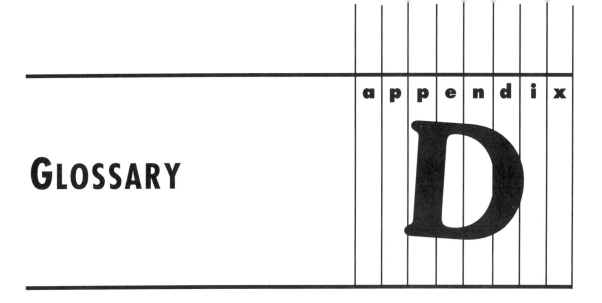

GLOSSARY

23B+D

The Primary Rate Interface (PRI) in ISDN. A circuit with a wide range of frequencies that is divided into 23 64 Kbps "bearer" channels for carrying voice, data, video or other information simultaneously and one D "delta" 16 Kbps channel for telephony data. See also PRI.

2B+D

The Basic Rate Interface (BRI) in ISDN. A single ISDN circuit divided into two 64 kilobits per second (Kbps) digital B channels for voice or data and one 16 Kbps D channel for low-speed data and signaling. Either or both of the 64 Kbps channels may be used for voice or data. In ISDN, 2B+D is carried on one or two pairs of wires (depending on the interface). See also BRI.

5ESS

A digital central office switching system made by AT&T.

Analog

Most current residential telephone lines are analog. Technically it is an electrical circuit that is represented by means of continuous, variable physical quantities (such as voltages and frequencies), as opposed to discrete representations (such as the 0/1, off/on representation of digital circuits).

B Channel

A "bearer" channel is a fundamental component of ISDN interfaces. It carries 64 Kbps in either direction, is circuit switched, and can carry either voice or data. See also BRI and ISDN.

BONDing

Bandwidth ON Demand (sometimes written BONDING) is the automatic combining of both B channels into a 128 Kbps channel for faster data transfer.

BRI (Basic Rate Interface)

One of the two types of interfaces in ISDN. BRI consists of two bearer or B channels and one data or D channel. Each B channel is 64 Kbps "clear" of digital bandwidth, which means that the full 64 Kbps is available to your application. The B channel carries the traffic. Any network control signaling is done externally to the B channel. The D channel is a 16 Kbps packet switching circuit. The network control signals are transmitted over this circuit. End-user applications can also use the D channel for low-speed data-only transmissions. One BRI standard is the U interface, which uses two wires. Another BRI standard is the S/T interface, which uses four wires.

CCS (Common Channel Signaling)

An integral part of ISDN known as "Signaling System 7," CCS is a method for sending call-related information between switching systems by means of a dedicated signaling channel. This signaling channel is separate from the bearer or B channels. CCS allows services such as call forwarding and call waiting to be provided anywhere in the network. Other acronyms for Common Channel Signaling are CCSS, CCSS7, and SS7.

Centrex

Centrex is a type of business telephone service. It is like having a PBX located in your local central office. Centrex is basically single-line telephone service delivered to individual desks (the same as you get at your house) with additional features.

CO (Central Office)

The central office (or CO) is a facility that serves local telephone subscribers. In the CO, subscribers' lines are joined to switching equipment that allows them to connect to each other for both local and long distance calls.

CPE (Customer Provided Equipment, Customer Premise Equipment)

Telephone equipment key systems, PBXs, NT-1s, answering machines, etc., that reside on the customer's premises.

D channel
> In an ISDN interface, the data or D channel is used to carry control signals and customer call data in a packet switched mode. In the BRI (Basic Rate Interface), the D channel operates at 16 Kbps, part of which will handle setup, teardown, and other characteristics of the call. 9600 bps will be free for a separate conversation by the user. In the PRI (Primary Rate Interface), the D channel runs at 64 Kbps. The D channel is sometimes referred to as the delta channel. See also BRI, PRI, and ISDN.

DMS100
> A digital central office switching system made by Northern Telecom.

EKTS (Electronic Key Telephone Service)
> This is a service that provides PBX-like capabilities using ISDN add-on features. It ties these add-on features to keys on your ISDN telephone, allowing you to have a hold button or a forward button, for example.

FPS (Frames Per Second)
> This term is most often used when talking about the speed of video capture and playback. It refers to the number of frames per second of video images displayed on the screen. The higher the frame rate, the more fluid the motion appears. The highest, or best, quality frame rate available is 30 fps. Lower frame rates (below 10) still appear as motion, but are noticeably "jerky," while zero fps is a still frame (no motion).

Gbps (Gigabits per second)
> A measure of bandwidth or throughput, based on 2^{30} (1,073,741,824) bits per second (slightly over a billion).

IEC
> See *IXC*.

ISDN (Integrated Services Digital Network)
> ISDN is the telephone network using all-digital rather than analog signals. It is a type of telephone service in which voice, data, and video information is digitized and transmitted at high speed over a single, public switched network. With ISDN, existing switches and wires (in most cases) are upgraded so that the basic "call" is a 64 Kbps end-to-end channel. ISDN comes in two basic flavors: Basic Rate Interface (or BRI) and Primary Rate Interface (or PRI). See also BRI and PRI.

ISP (Internet Service Provider)
An organization that provides access to the Internet, typically for a fee (determined by the bandwidth and availability of the connection provided).

IXC (Inter eXchange Carrier)
Also referred to as IEC. Any common carrier authorized by the FCC to carry customer transmissions between LATAs. AT&T, MCI, and Sprint are examples of Inter Exchange Carriers.

Kbps (kilobits per second, sometimes written as Kbps)
A measure of bandwidth or throughput, based on 2^{10} (1024) bits per second as a measure.

LATA (Local Access and Transport Area)
A geographic territory used primarily by local telephone companies to determine charges for intrastate calls. As a result of the Bell divestiture, switched calls that both begin and end at points within the LATA (intraLATA) are generally the sole responsibility of the local telephone company, while calls that cross outside the LATA (interLATA) are passed on to an Inter eXchange Carrier (IXC).

LEC (Local Exchange Carrier)
The local phone companies—either a Regional Bell Operating Company (RBOC) or an independent phone company (e.g. GTE)—that provide local transmission services. See also RBOC.

Local loop
The physical wires that run from the subscriber's telephone equipment to the switching system in the telephone company's central office.

Loop qual (Local loop qualification)
This is the process of checking the local loop distance—the distance between the customer's telephone equipment and the central office switch. If this distance exceeds 18,000 feet, additional equipment, such as repeaters, are added to the line to boost or enhance the signal. See also Repeater.

Mbps (megabits per second)
A unit of bandwidth or throughput based on 2^{20} (1,048,576) bits per second (a little over a million).

MUX (multiplexer)
An electronic device for combining multiple data or voice signals into one signal group for transmission over a high-speed trunk.

NI1 (National ISDN 1)

National ISDN 1 is a specification for a "standard" ISDN phone line. The goal is for National ISDN 1 to become a set of standards which every manufacturer can conform to. For example, ISDN phones that conform to the National ISDN 1 standard will work, regardless of the central office the customer is connected to. Future standards, denoted as NI2 and NI3, are currently being developed.

NT1 (Network Termination type 1)

The NT1 is the classic ISDN "black box." The NT1 is a customer premise device that converts the two-wire line (or "U" interface) coming from your telephone company to a four-wire line (or "S/T" interface). The NT1 connects between the ISDN phone line and the terminal adapter (such as the ISDN board in an Intel ProShare Video System). It supports network maintenance functions such as loop testing. as many as eight terminal devices may be addressed by an NT1.

NT2 (ISDN, Network Termination type 2)

An intelligent customer premise device, such as a digital PBX, that can perform switching and concentration. See also NT1.

Packet switching

Sending data in packets through a network to some remote location. The data to be sent is subdivided into individual packets of data, each having a unique identification and each carrying its destination address. Each packet can go by a different route and may arrive in a different order than it was sent. The packet ID lets the data be reassembled in proper sequence.

PBX (Private Branch Exchange)

A PBX is a private telephone switch. It is connected to groups of lines from one or more central offices and to all of the telephones at the location served by the PBX.

POP (Point of Presence)

A long-distance carrier's office in your local community. A POP is the place where your long-distance carrier, or IXC, terminates your long-distance lines just before those lines are connected to your local phone company's lines or to your own direct hookup. Each IXC can have multiple POPs within one LATA. All long-distance phone connections go through the POPs.

POTS (Plain Old Telephone Service)
The basic analog telephone service: standard single-line telephones, telephone lines, and access to the public switched network.

PRI (Primary Rate Interface)
The Primary Rate Interface (that which is delivered to the customer's premises) provides 23B+D or 30B+D running at 1.544 megabits per second and 2.048 megabits per second, respectively. See PBX.

RBOC (Regional Bell Operating Company)
One of seven regional telephone companies created by the AT&T divestiture: Nynex, Bell Atlantic, Bell South, Southwestern Bell, U.S. West, Pacific Telesis, and Ameritech.

Repeater
Equipment that is inserted at some point along the transmission line to amplify the signal. Repeaters are often used to "boost" a signal traveling over long distances, as in the case of lines that fail the local loop qualification test. See also Loop qual and Local loop.

Restricted Service Line
An ISDN trunk capable of operating at 56 Kbps. Restricted service line is common in the United States and Japan.

Service Order Number (SON)
The number used by the Local Exchange Carrier (LEC) to track your ISDN order.

SPID (Service Profile IDentifier)
The SPID number is issued by the telephone company and is used to identify your ISDN line to the central office switch. The Intel ProShare Video installation software prompts you for your SPID number, so it is important that you have this number on hand.

S/T Interface
A four-wire ISDN circuit. The more common ISDN circuit is the U interface.

SLC (Subscriber-Loop Carrier)
A computerized substation of the phone company for ISDN outside the 3.4 mile range of the central office switch; used instead of a repeater.

SON (Service Order Number)
The number used by the Local Exchange Carrier (LEC) to track your ISDN order.

SVN (Subscriber Verification Number)
> The number used by the Inter eXchange Carrier (IXC) to track your ISDN order.

T1 (also sometimes written T 1)
> A 1.544 megabit per second transmission method.

TE (ISDN Terminal Equipment)
> The equipment that is attached to the end of the ISDN line—an Intel ProShare Video System 200, for example.

TE1 (ISDN Terminal Equipment type 1)
> ISDN terminal equipment designed to connect directly to the S/T interface.

TE2 (ISDN Terminal Equipment type 2)
> Any equipment that has the ability to connect to the S/T interface for control/communications purposes via an ISDN terminal adapter.

Trunk
> A communication channel between switching devices or central offices.

U interface
> A two-wire ISDN circuit-essentially today's standard one pair telephone company local loop made of twisted-wire. The U interface is the most common ISDN interface.

Unrestricted service line
> An ISDN trunk capable of operating at 64 Kbps. Unrestricted service line is standard in Europe; some are available in the United States.

Index

Numerics
23B + D 27, 291
2B+D 26, 291
3Com 131
5ESS 291

A
A bit 55
AAL layer 77
ACC Nile 154
access links (A-links) 80
access unit (AU) 44, 70
acknowledged LAPD operation 63
acknowledged transfer 60
acronyms 204
ADAK 134
adapter card 99, 105
address 32, 60, 61, 245
ADTRAN 136, 149
Alpha Telecom, Inc. 149
analog 12, 291
    signal transmission 16
    telephone line 254
Ascend 155
asynchronous transfer mode (ATM) 74
AT&T 150, 191, 192

ATM 45, 87, 256
    adaptation layers 75
    layer 75, 76
automatic number identification (ANI) 84
availability 3

B
B channel 291
B8ZS 58
bandwidth 16, 24, 90
basic rate interface (BRI) 26
Bay Networks 156
BBSs 254
B-channel 24, 25
bearer service 68
    attributes 31
Bellcore 209, 263
bibliographies 288
B-ISDN 24, 25, 74, 86, 256
    model 88
bit stuffing 61
BONDed B channel 98
BONDing 257, 292
BRI 31, 253, 292
    frame 56
bridge 40, 99, 153

bridge links (B-links) 80
broadband 74
    services and standards 89
Broadband ISDN (B-ISDN) 25
business office 103

C

cables 4
    coaxial 39
    fiber optic 39
call reference flag 67
call setup 69, 262
Caller ID 84, 255
CCITT 20
CCS 79, 292
    network 80
central offices (COs) 13
Centrex 42, 292
    and ISDN 42
    LANs 44
channel service unit (CSU) 39
channels 23, 49
Chase Research 108
checklists 206
choices 105, 203
circuit mode calls 68, 70
circuit switched 31
Cisco 110, 157
CO 292
coaxial cable 39
Combinet 112, 158, 159, 179
commercial software 170
common channel signaling (CCS) 79
communications
    analog 12
  basics 12
    digital 14
      services 89
CompuServe 278, 288
configuration 53, 218, 235
conflict, hardware 245
connections 31, 53

consultant 6, 101, 204, 205
contention 56
control 41, 60
convergence sublayer (CS) 77
cost 5
    installation 97
    T1 40
country code 32
CPE 292
C-plane 30, 49
cross links (C-links) 80
CU-SeeMe 199
custom local area signaling services
    (CLASS) 84
customer premises equipment (CPE) 39

D

D channel 293
data exchange interface (DXI) 78
data link connection identifier (DLCI) 65
data link layer, 2, 30
data service unit (DSU) 39
data-based services 85
D-channel 24, 59
    Layer 3 protocol 66
    signaling 66
DEC 183
devices 27, 52, 253
    European ISDN 260
dial-tone service providers 263
DigiBoard 114
digital
    communications 14
    first carrier systems (T1) 15
    signal transmission 16
disconnect 69
distribution services 89
DLL 168
DMS100 293
DS1 57
DUP 83

E

E bits 55
EKTS 210, 293
Electronic Mail Essentials xvi
Ethernet 43, 100
Euronis 116
European ISDN devices 260
evolution 11

F

F bit 55
FA bit 55
FCS 62
features 211
fiber optic cable 39
flag 60, 61
FPS 293
frame 54, 57
    alignment sequence (FAS) 57
    check sequence 58, 60
    relay 41, 45
    mode 71, 74
FTP 213
full-motion video 257

G

Gandalf 160
Gbps 293
general 288
glossary 291

H

hardware conflicts 245
H-channel 24, 25, 58
HDLC 59
HDTV 89
home 215
    wiring 218
    office 102

I

IBM 137, 150, 193, 198
IEC 293
IETF 45
I-frame 62
INAR 172
in-band signaling 79
infrared 39
installation 4, 100, 218, 220, 235
    costs 97
integrated digital network (IDN) 19
Integrated Services Digital Network *see* ISDN
integrated voice/data LAN (IVDLAN) 44
Intel 117, 198
interface card, purchasing 217, 233
internal adapter card 107
International Telegraph and Telephone
    Consultative Committee (CCITT) 20
Internet 45, 96, 281, 289
    connection 228
Internet access software 213
Internet Protocol (IP) 45
Internet Service Providers 98
interworking 33
invoke 71
IP 24
IRQ 245
ISDI 170
ISDN (Integrated Services Digital
    Network) 288, 293
    bridges and routers 105, 153
    Ethernet bridge or router software 169
    hardware 105
    Internet service providers 212
    LANs 43
    "modem" 99, 105, 129
    needs 95
    PBX 42
    phones
    cost 255
    planes 49
    protocol architecture 49
    provider 209
    software 167
    terminals 68
    virtual circuits 70

ISDN Systems, Inc. 118, 194
ISDN*tek 120
ISP 98, 228, 294
ISPA 171
IVDLAN 44
IVDM 44
IVDTE 44
IXC 294

K
Kbps 294

L
L bits 55
LANs 18
    and PBXs 43
LAPB 66, 72
LAP-D 30, 31
LAPD 59
LAPD
    channel priority 65
    frame structure 60
    multiplexing 64
LATA 294
Layer 1 30
Layer 2 30
Layer 3 30
layers 252
LEC 294
LIDB 85
lightning 161
line problems 243
link access procedures 30
Lion 124
local access and transport areas (LATAs) 14
local area network (LAN) 14
local area networks (LANs) 43
local exchange (LE) 24, 64
local loop 23, 294
long-distance carrier 265
loop qual (local loop qualification) 294

M
Macintosh 100, 182
mail 213
mailing lists 284
MBONE 199
Mbps 294
message transfer part (MTP) 81
Microsoft 174, 181
microwave 39
Modem (MOdulate/DEModulate) 18
monthly service fees 97
Motorola 141, 143, 151
M-plane 30, 50
MPP 168
µ-law encoding 16
multiframes 57
multiplexers 17, 40
MUX 294

N
N bit 55
narrowband 25
national destination code 32
National ISDN 255
    HotLine 209
NetManage 176
network 13
Network Design Essentials xvi
Network Express 162
network Layer 3, 30
news 213
newsgroups 284
NI1 295
NIUF 255
NLPID 46
NT 54
NT1 27, 28, 99, 131, 148, 253, 295
NT2 27, 28, 295

O
on-line 276
ordering ISDN service 216–217, 232, 234

OSI 30, 50
    Layer 1 30, 51
    layers 30

P

packet mode connections 70
packet switched 31, 295
    transmissions 24
passband 16
PBX 103, 295
    and ISDN 42
    and LANs 43
    connections 41
    devices 41
    networks 41
PC adapter card software 168
personal use 101
physical layer 30
    protocols 51
physical medium (PM) 75
PlanetPPP 182
POP 295
POTS 54, 296
power 54
power supply 54
PPP 98, 168
PRI 57, 192, 296
Primary Rate Incorporated 196
primary rate interface (PRI) 26
priority classes 65
private branch exchanges (PBXs) 40
problems
    hardware conflicts 245
    line 243
    Windows for Workgroups 246
    Windows NT 246
protocol discriminator (Q.931) 67
protocols 29
    TCP/IP 45
pseudoternary coding 54, 65
purchasing your ISDN TA/NT1 interface card 217, 233

R

R reference 29
Racal 145
RBOC 97, 209, 296
reference point 29, 52
Regional Bell Operating Companies (RBOCs) 9
repeater 296
request for proposal (RFP) 205
requirements 90
resources 288
restricted Service Line 296
RFC 45
    1294 46
    1356 46
router 40, 99, 153

S

S bit 55
S interface 42
S reference 29
S/T interface 296
SAPI 64
satellite 39
scalable 87
SCP 80
search engines 282
segmentation and reassembly sublayer (SAR) 77
service access point identifier (SAPI) 61
service line
    unrestricted 297
service order number (SON) 296
services, communications 89
setup 68
S-frame 62
shareware 169
signal transmission
    analog 16
    digital 16
signaling connection control part (SCCP) 81, 82
signaling system no. 7 (SS7) 78

single-stage addressing 33
SIP 78
SLC 296
SLIP 98
SON 296
SP 80
specifications, technical 49
SPID 259, 296
SPIDs 4
SRAM 245
SS7 78
    protocol 66, 81
    services 83
standards 19
STP 80
structures 23
subaddress 32
subscriber loop 14, 23
subscriber number 32
subscriber-to-network interface (SNI) 77
Sun 199
Surf Communications, Inc. 126
SVN 297
switched 56 261
switched multimegabit data service
    (SMDS) 77
switching 15, 17

T
T reference 29
T1 15, 297
    costs 40
    hardware requirements 39
    networks 38
TA 27, 28, 99
TCAP 83
TCP/IP 45
    stack 167
TE 297
TE1 27, 28, 297
TE2 27, 28, 297
technical overview 9
technical specifications 49

TEI 64, 259
Telebit 163
teleconferencing 2
TeleMark 200
telephones 189, 190
telephony 12
terminal adapters 261
terminal endpoint (TE) 64
terminal endpoint identifier (TEI) 61
test equipment 201, 249
testing 218, 235, 243
time division multiplexing (TDM) 17, 24
timing 58
tips 243, 247
T-plane 30, 50
transmission 15
transmission convergence (TC) 75
troubleshooting 243
trunk 297
TSAPI 43
TUP 83
two-stage addressing 33

U
U interface 57, 58, 297
U reference 29
U-frame 62
UI frame 64
unacknowledged LAPD operation 64
unacknowledged transfer 60
unrestricted service line 297
U-plane 30, 50
user and application part 81
user-network signaling 66

V
vendor 267
video, full-motion 257
video conferencing 197
virtual circuit 73
VMS 183
VSAT Ku-Band 199

W

WANs 18
wide-area network (WAN) 14
Windows 100
Windows for Workgroups, solutions 246
Windows NT 181
    solutions 246
WINSOCK.DLL 168
wireless and satellite equipment 199
wiring 4, 218, 235
    home 218
WWW 96, 108, 213

X

X.25 24, 66, 72, 73
X.25 packet service 41, 45

Z

ZyXEL 145